iPad Music

iPad Music

In the Studio and on Stage

Mark Jenkins

Focal Press
Taylor & Francis Group

NEW YORK AND LONDON

First published 2013
by Focal Press
70 Blanchard Rd Suite 402 Burlington, MA 01803

Simultaneously published in the UK
by Focal Press
2 Park Square, Milton Park, Abingdon, Oxon OX14 4RN

Focal Press is an imprint of the Taylor & Francis Group, an informa business

© 2013 Taylor & Francis

Notices
Knowledge and best practice in this field are constantly changing. As new research and experience broaden our understanding, changes in research methods, professional practices, or medical treatment may become necessary.

Practitioners and researchers must always rely on their own experience and knowledge in evaluating and using any information, methods, compounds, or experiments described herein. In using such information or methods they should be mindful of their own safety and the safety of others, including parties for whom they have a professional responsibility.

Product or corporate names may be trademarks or registered trademarks, and are used only for identification and explanation without intent to infringe.

Library of Congress Cataloging-in-Publication Data
Jenkins, Mark, 1960–
    iPad music: in the studio and on stage/Mark Jenkins.
        p. cm.
1. iPad (Computer) 2. Computer sound processing. 3. Computer music—Instruction and study. I. Title.
        ML74.4.I477J46 2013
        781.3'45365—dc23

ISBN: 978-0-415-65680-1 (pbk)
ISBN: 978-0-203-06983-7 (ebk)

Typeset in Berling
Project Managed and Typeset by: diacriTech

SUSTAINABLE
FORESTRY
INITIATIVE

Certified Sourcing
www.sfiprogram.org
SFI-00555
The SFI label applies to the text stock.

Printed and bound in the United States of America by Walsworth Publishing Company, Marceline, MO.

Table of Contents

Introduction & Notes

Apple's iPad has been the realisation of a long-held dream for many musicians, offering a portable, powerful, versatile and great-sounding electronic instrument that interfaces easily between existing home and studio recording equipment.

All this was far from being one of the iPad's original main purposes – in fact, the device was launched in April 2010 without even offering any version of Apple's free music creation software GarageBand and was prevented in many ways from transferring musical or other data as any average laptop computer could have done. But the unrealised potential of the device was very quickly spotted by a horde of mainly third-party developers, and at the time of writing, a huge choice of musical apps and hardware add-ons has become available.

Ironically, the iPad was far from the first attempt to launch a mass-market tablet device (Apple's own first effort being the short-lived Newton in 1993) and was not even a complete innovator in the field of tablet music. As early as 2003, companies already offering tablet PC computers included Compaq, and their TC1000 model marketed through Hewlett Packard offered a stylus-operated tablet fully detach-able from its keyboard and was very successfully used by computer-based musicians including Todd Rundgren.

The Compaq models remained a novelty, and attempts to launch mass-market tablet devices by companies such as US Robotics with

Music technology expert MARK JENKINS performing at the GADGET SHOW LIVE at the huge National Exhibition Centre near Birmingham. Mark featured on the show's current season as the first musician to release an album of all-iPad music, and has been giving live workshops on creating music for the iPad, smartphones and Android tablets.

www.markjenkins.co

FIGURE 1   Mark Jenkins Gadget Show Live

the Palm Pilot had met with very limited success. On the whole, since laptop computers were already becoming smaller and more powerful every month, it seemed the public did not see the purpose of a very small tablet device with potentially very restricted abilities.

Why the iPad took off so spectacularly on its launch in April 2010 remains a subject for debate. Apple's iPhone had already made a huge impact, and the market was keenly anticipating a device somewhere in size and abilities between the iPhone and the Macintosh range of laptop computers. However, on launch it appeared that the iPad would cost almost as much as a laptop computer and would offer very few of the storage and expandability options (lacking any hard disk drive or USB port) while also omitting the telephone facilities of the iPhone.

Media applications were also limited. The iPad supported a version of iTunes for music playback, but had no independent video output and could not send its full screen display to external monitors, other than when using the Video Player and a handful of other applications. There was no version of iMovie available, so possible use as a video editing device seemed unlikely, and as mentioned, musicians already well used to composing and creating original music with Apple laptops as well as with PC's were disappointed by the omission of any version of GarageBand, Apple's free music package included with the OSX operating system – let alone the more advanced Logic Express or Logic Studio packages, or anything resembling competing sequencing software, such as Cubase or SONAR.

Initially, the solution to these limitations lay in the very rapid growth of the Apps Store, Apple's online resource for new free or purchased iPad software. Opening the App Store up to third-party developers, many of whom had already been writing extensively for the iPhone, led to an explosion of iPad apps which very quickly transcended the limitations of the device as launched. Within a matter of weeks, the true potential of the iPad for musicians, as for many other users, began to become obvious.

The iPhone had already been offering a few music-related apps, but thanks to its very small display and limited processing power, these had been limited to acting mostly as musical toys. However, when some of these were very promptly extended to take full advantage of the iPad's much greater screen size and processing power, a potentially very flexible and versatile professional standard musical instrument was very quickly born.

The following are a few notes before continuing to the rest of the content of this book.

1   iPad models

All the hardware products discussed in this book will work for all models of iPad. All apps will also run on the original iPad, iPad 2 and third-generation iPad regardless of memory size. Just a handful of Apple app products such as iMovie will not run on the original iPad, supposedly because it lacks a built-in camera.

There are instances where apps will perform slightly better on the later model iPads, thanks to faster processing speeds – for example, the Sunrizer synth app produces 8-voice polyphony on an original iPad and 12-voice polyphony on an iPad 2. But these variations are very minor, and musicians interested in getting into the use of the iPad can certainly consider a second user market original model iPad with perfect confidence, particularly since updates to the iOS used to run the iPad and iPhone will load and run on all model iPads.

If a smaller iPad model probably in a seven-inch format appears, it's very unlikely that it will be anything other than completely compatible with the existing iOS, though some manufacturers of music hardware would have to retool to accommodate a different size of casing.

2   iPhone versions

Many of the hardware products discussed will work for the iPhone (usually the 4/4S models) and sometimes for the iPod Touch as well as for the iPad, as long as their function depends on interfacing with the audio and dock ports of the iPad rather than on accommodating it physically.

Many of the iPad apps discussed have versions that will run on the iPhone (and sometimes on the iPod Touch too), so this will not always be mentioned specifically – the App store always makes very clear which apps have iPhone versions available. In almost all cases, iPhone versions of apps will be considerably simplified, not least due to the smaller screen display. Virtual keyboards will typically be smaller, synthesizer apps will typically offer less polyphony, drum sound apps will offer fewer virtual pads and so on.

When purchasing an iPad app from the App Store, the iPhone version if available generally becomes accessible too, so if you get used to working with a package such as GarageBand on the iPad, you may possibly get some benefit out of having the even more portable iPhone version available too. Certainly, many iPad apps along the lines of guitar multi-effects are readily available to run on the iPhone, which can then do a great job as a miniature multi-effects device.

3   Prices

All prices are stated in US dollars and are approximate full retail prices at the time of writing. Online prices from reliable sources such as Amazon will often be lower. UK prices in pounds at the time of writing are around 60% of those quoted in dollars.

Chapter 1

Computers for Music

The creation of music using electronic instruments, and later computers, laptops and tablets, dates back to the 19th century, and a vast amount of extra detail is available in my book *Analog Synthesizers: Understanding, Performing, Buying*. This charts the development of electrical and electronic (often valve-based) instruments such as the Telharmonium, Ondes Martenot and Hammond organ, through Dr. Robert Moog's creation of the familiar analog synthesizer based on a number of now well-understood components or modules, such as audio oscillators, filters and amplifiers, to the simulation first of analog textures and later of much more complex systems of sound creation within computer software packages.

Using a computer to create music has been of interest from the very first generations of digital equipment, such as the RCA computer and the DEC PDP8, through the early Apples and PCs, the MIDI-equipped Atari ST and towards modern laptops. Whether the computer itself was carrying out any sound creation, in addition to controlling or playing notes on other instruments usually connected through a MIDI or other type of interface, would always depend on the quality of sounds required and the processing power available to the computer. The first on-board home computer sounds, such as those generated by the SID audio chip in the Commodore 64 computer, were poor examples indeed, but the quality of sounds available increased precipitously, to the point where early music software such as ReBirth could very effectively substitute for hardware instruments such as Roland TB-303 Bassline synth and Drumatix drum machine.

With the floodgates open for computers and laptops to actually generate sounds as well as to control them, the quality and variety of sounds available could only increase. Around 2004, I recorded my final CD album using a huge collection of around 50 analog and MIDI synthesizer keyboards and modules built up over a period of years and began work on a new release, 'If the World Were Turned On its Head, We Would Walk Among the Stars', using only virtual or 'soft' synthesizers running on a G4 tower Macintosh. Very shortly after this time, it became possible to run a reasonable number of synthesizer voices with professional sound quality on a laptop, and I recorded and performed both in the United Kingdom and in the United States using a MacBook laptop running Steinberg Cubase together with a large number of soft synths. At the time of the release of the iPad, I had just completed a techno/ambient album, 'Something Dancing in the Darkness', under the band name, Ghosts of Mars, using only Apple GarageBand and its built-in software synthesizers running on a MacBook Pro.

This continuing move towards simplification, the removal of a dependence on external instruments or modules and a wish to perform more widely in the United States and around the world as well as in the United Kingdom and Europe, made a musically capable iPad an extremely attractive proposition for me, as for many other musicians around the world. However, initial reports made the iPad appear much limited for any such use, with no music creation software apps available on its initial release. However, taking the plunge with a 64 GB original model iPad, it was only a matter of weeks before App Store releases began to appear, which seemed to give the iPad some considerable potential for music creation.

Chief amongst these apps was iSequence, from the independent Polish designers Beep Street. Like all App Store releases, it appeared at what seemed (for anyone already used to buying Mac software packages and sequencing software such as Cubase or Logic) an astonishingly affordable price of a few dollars, and yet seemed incredibly able. iSequence offered sequence composition, extensive song capacity, built-in synthesizer sounds, built-in drum sounds, versatile mixing and built-in effects, in fact all that was needed to create complete pieces of instrumental music (although not so much in the way of songs or anything else requiring acoustic or electric instruments such as guitars recorded – a matter which we'll investigate later on).

To analyse the power, which iSequence running on the iPad appeared to offer at that time, let's look at the styles of music with which I'd been involved since the birth of modern electronic music in the 1960s.

While the 1950s was mostly about the tape manipulation of Karlheinz Stockhausen and Pierre Schaeffer, the late 1960s and early 1970s saw the birth of the Moog synthesizer and its adoption by progressive rock musicians such as Keith Emerson with Emerson, Lake & Palmer, jazz performers such as Jan Hammer and Chick Corea and the more experimental German musicians such as Klaus Schulze, Kraftwerk and most notably Tangerine Dream. More than those in other countries, the German synthesizer musicians also adopted the analog sequencer as part of their instrumentation, and so their compositions often included very metronomic, perfectly timed, rather minimalist or repetitive elements.

The key to the artistic success of these performers lay in combining these automated rhythms with much more expressive melody, harmony and highly imaginative textural parts – which is why their later, much less musically capable and tasteful imitators, fell so far short of their achievements. However, one musician who successfully brought these styles to a mass market was Jean-Michel Jarre, and his multi-layered highly evocative synthesizer textures remain popular to date.

What all these performers had in common was a requirement for a vast amount of instrumentation, to play repetitive sequencer parts, percussion tracks, harmony and textural parts, sound effects and melodies with a great variety of style and texture. So Jean-Michel Jarre in concert, for example, would typically be surrounded by a Moog Modular synthesizer and custom-built sequencer, several drum machines, a Yamaha CS-80 or later on the bulky Memorymoog for chords, an ARP 2600 semi-modular synth and an EMS VCS for sound effects, an Eminent organ for strings, a Theremin for 1950s sci-fi style swooping effects and maybe a portable controller keyboard, drum pads and more. That's a pretty substantial stage setup, generating a requirement for large stages on which to perform a road crew, a tour bus and so on, hardly a quick and spontaneous way of creating electronic music. Similarly, the Berlin studio of Tangerine Dream in the 1980s would have been filled with MIDI modules and controlling computers, plus keyboards from Moog, the German manufacturer PPG, Roland, Korg, audio samplers from Akai and instruments from dozens of other manufacturers.

Replacing a whole stage or studio full of keyboard instruments, or a controlling computer plus a whole rack of MIDI modules and samplers, with a single small self-contained instrument which could produce everything from repetitive sequence and drum parts to lush harmonised textures, melodies and sound effects would seem to be an outrageous demand, and certainly, the iPad when launched wouldn't appear to have been up to the task. However, the earliest version of the iSequence app went an astonishingly long way towards realising this ambition. Here were 8 tracks of 32 events pattern, each track able to play a synthesizer or drum sound chosen from an onboard selection of several hundred, the 32-event patterns capable of being edited and chained into complete songs, coupled with an 8-channel mixer with mute and solo abilities, plus 3 effects including delays and reverb which could be assigned to any track, and a filing system which meant that songs could be loaded and played quickly even under stage conditions. This seemed a more than capable system to attempt the creation of pieces roughly in the style of the musicians mentioned above – pieces which relied a great deal on sequenced repetition rather than on recorded vocals or on acoustic instruments.

FIGURE 1 The iPad Album

After the appearance of iSequence, a few weeks after the iPad's launch, I started working on a complete set of pieces using only iSequence running on a single iPad and released a CD 'The iPad Album', first performed in Holland in October 2010 and commercially available ever since, including through Amazon worldwide. This was the world's first all-iPad commercial CD, predating by some months a free download album by Gorillaz (who apparently didn't know of iSequence), who used an iPad for some basic sketches before songs were completed with guitars, voices and other instrumentation in a large multi-track studio, and much later use of the iPad by Björk and other musicians.

One of the advantages of iSequence is that it has a virtual keyboard as part of its screen display, so notes can be entered in step time or (with care) in real time, without the need to connect any external keyboard. While laptop users would have been used to connecting MIDI or USB keyboards by this time, the iPad in fact didn't offer this facility at all

Mark Jenkins "The iPad Album"
Launch concerts   Holland 9th October 2010   www.markjenkins.co

FIGURE 2   Mark Jenkins iPad Concert

on its initial launch. It's sole audio socket appeared to offer audio out only (in fact it does also offer mono audio in, as we'll see shortly) and the only other connector was the Apple dock, for which no musically oriented add-ons were available at all.

The history of how this situation rapidly changed comprises the remaining content of this book. Although there are many fascinating and powerful music creation apps now available for the iPad, we'll concentrate here rather on the hardware add-ons which have made it increasingly practical to use these apps in everyday studio and stage musical applications. The iPad is of limited use as a music synthesizer if it can't be played from an external music keyboard, as a sound sampler if it can't be fed sounds from the world around it and as a sequencer if it can't be connected up to some of the familiar instruments which musicians still have on hand in their studios.

We'll be looking at how the iPad, at first enticing but rather enigmatically isolated from the world around it, rapidly became, with the introduction of dedicated audio interfaces, MIDI interfaces, add-on keyboards, control surfaces, drum pads and many more hardware options, the powerful and flexible music resource today.

Chapter 2

Interfacing the iPad

As early adopters immediately realised, the Apple iPad, although heavily oriented towards displaying photos, lacked the SD card or USB slots, which would have made it easy to load and update them. The USB connector also had become the most common substitute for the MIDI interface generally found on electronic instruments for over 20 years; so without either USB or MIDI interfacing on offer, the iPad at first seemed firmly cut off from the world of musical instrumentation.

Initial hardware add-ons offered very little encouragement. Apple's Camera Connection Kit (CCK), released almost simultaneously with the original iPad and still available now, comprised two small white adapters for the dock connection, one offering a slot for SD or MMC cards taken from a camera and the other offering a USB socket so that the images could be transferred directly from a camera without removing its memory card.

It is the USB adapter which is of interest to us here, because it was very quickly appreciated – although not widely publicised at the time by Apple – that USB-equipped devices other than cameras could also potentially take advantage of the interface. This was first tested with QWERTY keyboards, and indeed, it was found that with no modification, many USB keyboards could be used via the USB adapter of Camera Connection Kit, for typing within iPad Mail, Notes and other apps. This provided a handy alternative to buy one of Apple's dock-equipped keyboards or a wireless Bluetooth keyboard (the iPad's Bluetooth incidentally supporting

FIGURE 1   Apple CCK

keyboard, headset and headphone connections, but not file transfer of any kind, a situation which persists at the time of writing but may at sometime be the subject of an operating system update).

Very rapidly, it was considered that some USB-equipped audio interfaces and microphones may also work via the Camera Connection Kit, and this was quickly tested and found to be the case. The concept of a high-quality USB-equipped microphone had only become common just before this time, many manufacturers realising that the requirement for a large audio interface just to connect a single microphone to a laptop via USB seemed unnecessary. Manufacturers of good quality USB-equipped microphones included Blue, Shure and Apogee, and it was quickly found that many of these USB microphone models worked readily with the iPad using the Camera Connection Kit.

The number of iPad apps which could actually take advantage of any external microphone remained strictly limited – remember that at launch, there was no version of Apple GarageBand and that iSequence handled only built-in sounds, not sampling or audio recording – however, some simple apps which recorded mono or stereo sound were found to work reliably, and now at the time of writing, there are very many more.

We'll look in more detail in Chapter 3 at some further microphones and audio interfaces, which work either through the iPad's audio connector or as discussed above, using the Camera Connection Kit. However, an even more significant development took place with the release of iOS 4.2 around November 2010, when it was realised (again without much discussion from Apple) that the USB/Camera Connection Kit/dock interface would now also process MIDI data. This was quickly tested with various music keyboards, synthesizers and apps, and many were found to work without any complication.

The significance for iPad music of MIDI interfacing via the Camera Connection Kit can't be overstated. Although the iPad's touch responsive screen is versatile, there's no substitute for being able to connect a conventional mechanical music keyboard as an input, while sending MIDI data out allows the iPad to act as a sequencer or controller for other instruments and to synchronise together time-dependent instruments, such as drum machine simulators, running on it.

The number of USB-equipped musical keyboards and other devices already on the market and not specifically related to the iPad is very large, to the extent that it is not practical to list them all – however, we'll mention some of those known to work particularly well with the iPad, in addition to cover the much smaller number of devices which do have specifically iPad-related connections or other design features.

Apple's Camera Connection Kit initially retailed at $29 and remains available. There are also imported compatible devices, some combining both the USB and the SD/MMC slot into a single device, and these have been tested and worked equally well in most cases. Having a handful of these, or of the original Apple devices, one attached to the output cable of each of your USB keyboards or to some spare USB cables, is by no means economically demanding.

Some MIDI/USB devices will fail to work directly with an iPad simply because the iPad does not supply enough current through the CCK to operate them. USB delivers 5 V of power, and some compatible devices will expect 1 A or even 2 A of current. However, an iPad delivers only 3.3 V from its dock port, and so the Camera Connection Kit does some DC to DC conversion up to 5 V, which limits the amount of current it can supply. USB outputs generally produce up to 500 mA, and so fairly large instruments – such as piano-sized controller keyboards with large illuminated displays – can be operated with no problem. Apple's suggested that maximum output current through the Camera Connection Kit is 10 mA, although some users appear to be able to run 20 mA or even 50 mA devices.

However, it is certain that some more demanding devices will simply not work, in which case the user would have to look at using a mains-powered USB splitter device. To get mobile again, it may be possible to run one of these from a portable battery box such as those marketed by Philips and other companies, through Radio Shack, Maplin and similar electronics stores.

When MIDI compatibility in iOS 4.2 was developed, any class-compliant and suitably low-powered MIDI USB device should have become compatible via the Camera Connection Kit. Based on user reports at the time, the following devices were found to work:

Akai LPD8 (small drum pad controller)
Akai LPK25 (miniature keyboard controller)
Alesis Q49 (controller keyboard)
Behringer BCF2000 (control surface)
Behringer BCR2000 (control surface)
Digitech GNX4 (audio and MIDI)
E-Mu XMIDI (2 × 2 MIDI interface)
Edirol PCR-M80 (MIDI interface)
Electron TM1 (drum machine)
ESI MIDIMate II (MIDI interface)
Korg NanoKey (miniature keyboard controller)
Korg NanoPad (miniature pad controller)
Korg NanoControl (miniature rotary controller)
Korg Triton TR series (keyboards)
M-Audio Axiom Pro 61 (keyboard)
M-Audio Pro Keys Sono 61 (keyboard)
M-Audio Uno 1x1 (MIDI interface)
Novation Remote 25LE (keyboard controller)
Novation Remote SL37 (MIDI controller)
Novation Xio 25 keyboard (audio and MIDI)
Sonuus i2M Musicport (MIDI interface)

A more extensive current list is given in the Appendices.

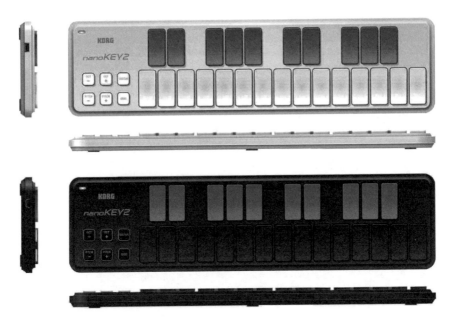

FIGURE 2 Korg NanoKey

Within a short time, the Apple's Camera Connection Kit and its clones were joined by other methods of interfacing audio and MIDI devices to the iPad. We'll look at the other audio options (which in some cases work through the audio connector) in Chapter 3 and at other MIDI interfaces working through the dock port in Chapter 4. Some more complex devices such as the Alesis i/ODock and TASCAM iU2 now offer both functions and are covered in Chapter 5.

Chapter 3

Audio Interfaces and Microphones

The nature of the iPad's audio socket is often not understood in detail, since it's almost exclusively used as a headphone or stereo audio output using a three-connector (TRS) 3.5 mm plug. On the original iPad, a built-in microphone was located immediately next to this socket, so it wasn't generally understood that the socket is in fact wired TRRS to support four connections, of which the last is a mono audio input intended for an alternative microphone. See the Appendices for more details.

The audio input is intended primarily for connection of an alternative microphone, but with a suitably wired source cable should also allow appropriately written apps to record audio from sources such as CD players (with mono signals only, not stereo). But since this sort of application would often favour stereo operation, a more complex interface using the more extensively enabled dock port is the more likely resource. The mono audio socket perhaps has a more obvious application in recording electric guitars – but these don't operate at mike level. Once this was appreciated, several products came onto the market that ensured the correct level input was offered together with, most often, the quarter inch mono jack socket typically required for connection via an electric guitar cable.

The simplest way to do this is using a suitably wired cable such as the $12 Musical Instrument Adapter from Peterson Phone Accessories. This features an inline jack socket on one end of a few inches of cable, a TRRS minijack plug for the iOS device's audio socket on the other end and allows guitars, electronic instruments and dynamic mikes to be connected to the iPad or iPhone's audio socket. But since it doesn't have a headphone output, you'll have to make your own arrangements to hear what you're doing. It's handy though if your iPad app of choice is simply a guitar tuner, and of course doesn't tie up the dock connector, so your iOS device can remain on charge.

FIGURE 1 MIDITech iPface

In Germany, the mail order specialist Thomann offers the Mididtech iPface that is almost as simple, offering a guitar input and audio connector output plus a headphone output, or the very similar Harley Benton Jackamp, each at around $20.

Equally simple at around $20 is the JamUp Plug, a white matchbox-sized device with a built-in TRRS plug plus a quarter inch guitar input and headphone output. It claims better crosstalk than some similar designs, and though it works with the GarageBand, AmpKit and AmpliTube apps, the designers also offer JamUp Lite and JamUp Pro apps that work well. With iTunes play-along, a tuner, metronome, overdub phrase sampler and many more functions, the apps have been praised as offering excellent value for money.

www.positivegrid.com

FIGURE 2   HB Jackamp

The limitations of any audio input device using the iPad's audio connector are as follows: First, it's mono (single channel) and not stereo.
Second, since the audio outputs and inputs are in very close proximity,
there's a possibility of noise and crosstalk that may even become as bad
as to create feedback. Third, the audio input is intended for use with
an alternative to the iPad's built-in microphone, and so it operates at
microphone levels as well as processing the sound to favour the quality
of the human voice. This doesn't matter so much for guitars, but for
bass guitars creates a significant cut in bass level.

Audio input devices using the audio connector must always be carefully chosen with these limitations in mind. Some are more carefully
designed than others to limit the possibility of crosstalk noise and feedback. Some designed to handle bass guitars will provide a good deal
of bass boost. The earliest guitar input devices very much relied on
the audio connector, and several companies had been developing their
own guitar effects software, first for the iPhone and later for the iPad.
Companies that offer a guitar-oriented audio input device sometimes
accompanied with own brand guitar multi-effects apps now include IK
Multimedia and Peavey.

If none of these relatively inexpensive audio input devices meet your requirements, it's time to look at the slightly more expensive audio input alternatives that use the dock connector.

Apogee's Jam at around $99 is one of the most straightforward hardware devices for guitarists wishing to interface with the iPad. Although Apogee don't offer any recording app of their own, Jam is readily compatible with Apple GarageBand and many other audio apps such as AmpKit Free.

FIGURE 3 Apogee Jam

The Jam device comprises a lipstick-sized tube with a quarter inch guitar socket at one end, rotary gain control on the side, a level indicator light on the top and a proprietary multi-pin socket on the top. Two short cables are provided to connect to this socket, one terminating in a USB plug for use with a computer and one in a dock connector for use with the iPad.

The advantage of Jam using the dock connector over input devices using the headphone socket is that the input is digital, and so not modified for the human voice, and that the audio output of the iPad remains available to be passed on to a larger amp or speakers. The disadvantage is that the iPad can't remain on charge while using the dock connector.

Apogee Jam works faultlessly and its single indicator light actually says more than you might imagine – changing colour according to whether you have a guitar connected but no software app (blue), ready to record (green) or overloading (red). I've also used the device for mono audio recordings from movie dialogue and other sources straight into iSequence and it works well, so can be recommended.

www.apogeedigital.com

Fractionally larger than the Jam is TASCAM's handy iXZ ($70), a single-channel mike, guitar or instrument input for the iPhone, iPad or iPod Touch, which takes two AA size batteries for a built-in mike preamp and to supply phantom power to condenser mikes (around 48 V or more typically 46 V at up to 5 mA, so you could easily use a vocal mike such as the Audio Technica AT3035).

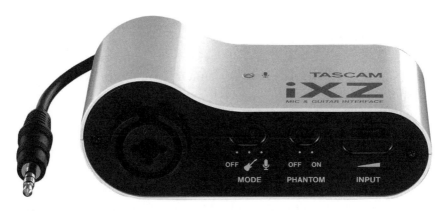

FIGURE 4   TASCAM iXZ

The iXZ connects to the iPad's audio input, leaving the dock connector free for charging and has a (very) short trailing lead to do so. On the front panel is a combined XLR/quarter inch jack socket, together with a sliding switch to select guitar, mike or line levels. Phantom power can be switched on or off and there's an overall input level control.

The rear panel features a minijack headphone socket for playback, and the device works with GarageBand, TASCAM's own four-channel Portastudio app, TwistedWave that can convert audio files to MP3 that isn't usually possible on an iPad and many other apps.

Many users have found the iXZ terrific value for money, but its overall noise performance will be limited by the lack of separation between audio input and output on the iPad's audio connector and by the combination of mikes and recording apps used. Experimentation is the key and the iXZ may be a good solution for field recordings, voiceovers and other work as well as for guitars and other single-channel instruments.

An alternative dock connector style audio input for the iPad is the Line 6 Mobile In ($80). We'll look in Chapter 4 at the compact MIDI Mobilizer from Line 6, and the audio-oriented Mobile In has the same tiny format with the dock connector on one side, and on the other side a stereo minijack Line Input and a mono minijack Guitar In for which a cable is supplied, only one of which can be used at any time.

Mobile In interfaces in a more integrated manner than usual with other products from Line 6, since the company has been making guitar effects, amps, digital guitars and effects software/apps for some time. Mobile POD is the appropriate free app from Line 6, only works if Mobile In is connected, and also allows firmware updates to be made. In GarageBand, you'll select Line In or Guitar In inputs, the Guitar input making the appropriate adjustments for impedance – though the device also works with guitar effects apps from other manufacturers.

The Mobile POD app provides 32 amplifier simulations, 16 stompbox and rack effects, 16 speaker cabinet simulations and 10,000 presets, so is ideal for recording direct into GarageBand or for playing along to iTunes tracks. Other apps with which the Mobile POD works are listed in www.line6.com/mobilein/apps, and another free app Jammit offer allows you to play along with remixed music tracks by bands including, for example, Rush (with guitar parts removed and a music score displayed), and record your efforts, automatically selecting suitable amp and effects settings for each song. Additional songs and effects have to be purchased.

Also using the dock connector, GuitarJack from Sonoma Wireworks ($150) is a small interface around half the size of an iPhone that, apart from a minijack headphone output, offers a quarter inch guitar input on one side and a minijack stereo mike/line audio input on the other side. Unusually, both can be used simultaneously, so you can record electric guitar and vocals together. Compatible apps include GuitarTone (free), FourTrack (iPhone) and StudioTrack (iPad) from the same company, as well as GarageBand. In the Sonoma apps, a control panel becomes accessible allowing the user to set input levels and modes including stereo or dual mono, so the device should work well with all sorts of audio inputs and mikes – although lacking battery power, it won't offer phantom power to condenser mikes.

In use, the GuitarJack works well, though some users complain about the fact that it cuts off the iPad's internal speaker, forcing the user to rely on headphones or external amplification and that adding a dock extension cable rather than having the device projecting directly from the iPad feels far more comfortable.

www.sonomawireworks.com

Much simpler and taking the form more or less of simply a cable, iRiffPort ($100) from Pocketlabworks has been gaining good reviews, other than for its dock connector that doesn't seem to lock very firmly. The 6 ft. cable, however, does offer a headphone socket at the dock end – in case you're standing pretty close to your iPad – and another at the guitar plug end, in case you're standing closer to your amps or other gear.

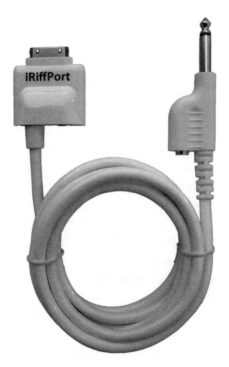

FIGURE 5 iRiffport

Other than those facilities, iRiffPort is pretty much featureless and comes accompanied by an app called PocketAmp that can be used in conjunction with GarageBand (it's best to start up PocketAmp first, then GarageBand afterwards). PocketLabWorks also market an extended app PocketAmp GK and a variation, RockShop, endorsed by Megadeth's David Ellefson.

iRiffPort gets generally good reviews and is sold in the United Kingdom by Absolute Music Solutions.

www.pocketlabworks.com

iRig from IK Multimedia ($40) is functionally identical to iRiffPort, taking the form of a lipstick-sized tube that accommodates a guitar, quarter inch jack plug, having on the opposite end a short captive cable to a TRRS plug for the iPad's audio socket, and a minijack socket for direct monitoring using headphones or amplifiers. iRig is not advertised as being suitable for line inputs other than guitar, so it's going to be matched most often with guitar multi-effects software such as IKM's own AmpliTube.

AmpliTube runs on iPad or iPhone and is one of the most sophisticated guitar effects app packages around – there's also a Fender version specifically modelling Fender amps and effects. In AmpliTube, a virtual set of stompbox effects pedals is matched with amp and cabinet models, and although the choices are limited in the free version, there are many more in the full version. Most users seem very happy with the iRig/AmpliTube combination. Some complain of noise or uncontrollable feedback, and as we've seen, noise performance and input/output separation using the audio socket on iOS devices is always going to be inferior to that when using the dock port.

But price is a consideration and with iRig available at some very affordable prices at the time of writing, it's going to be a first port of call for many who want to insert guitar signals into an iPad or iPhone.

iRig Stomp can be regarded as a version of the iRig input in footpedal form. It's a small black box featuring a chrome footswitch for true bypass and a large rotary gain setting control. Again, there's a guitar input and output to an iOS device as well as directly to a PA system and to headphones. iRig Stomp will sit happily in any footpedal setup and it's up to the player to decide how to configure its use alongside other pedals, guitar amps and cabinets.

See Chapter 7 for more on audio input devices specifically intended for the guitarist.

IKM also offers iRig Pre ($60), which does for microphones what iRig does for guitars. iRig Pre takes the form of a small rectangular black box accommodating a balanced XLR mike input at one end and a captive cable to a TRRS plug for the iPad or iPhone's audio socket, as well as a minijack headphone socket for direct monitoring, at the other end. Phantom power for the mike can be switched on or off and there's a rotary gain control on the side along with green and red input level LED's.

iRig Pre allows the user to match a favourite single-channel condenser mike to any iOS recording app and, of course, is well suited to use with IKM's own VocaLive app that offers a huge range of processing and effects.

www.ikmultimedia.com

A rather more sophisticated interface is offered by Studio 6 Digital, a company that offers an acoustical test app AudioTools for the iPad. Their iAudio Interface 2 ($400) comes in a sturdy metal chassis and offers line in and phantom powered XLR mike in together with line and headphone outputs and a captive dock connector cable. In fact, the company also offers a test mike – called iTestMic – with a dock connector built in, so you can bypass the use of an interface altogether.

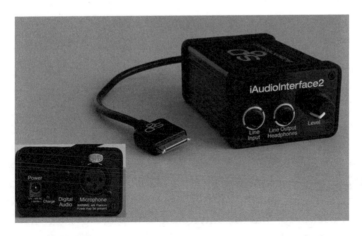

FIGURE 6 AudioInterface

www.studiosixdigital.com

MICROPHONES

As we've seen, USB-equipped microphones can generally be used with the iPad in conjunction with the Camera Connection Kit. There are now many of these on the market ranging from mono telecoms, voice-over and vocal mikes to high quality studio stereo condensers requiring phantom power and it's beyond the scope of this book to mention them all. A few example of manufacturers include MXL, Marshall and Samson who make the tiny GoMic with a laptop clip as well as the attractive Meteor desktop vocal or instrument mike, Shure (who also make an X2U interface to simply connect any existing XLR-equipped mike to USB), Chord, t.bone, Behringer, Logitech, Audio Technica and many others. If only one manufacturer could be mentioned in more detail, then the very affordable Yeti and even more compact and inexpensive Snowball (allegedly the world's first professional USB mike) both from Blue Microphones can be very highly recommended for their advanced specification and performance.

www.bluemic.com

As another example, Samson's Meteor ($150 or often a lot less) is a smart looking tabletop USB cardioid mike that, using the Camera Connection Kit, works well with GarageBand, FiRe and other recording apps.

www.samsontech.com

There are just a few microphone products that are more specifically dedicated to use with iOS devices.

TASCAM's iM2 at around $90 is one of the very few microphones specifically dedicated to use with iOS devices. It's tiny, just about the width of an iPhone (and finished in black or white to match either iPhone design), offering stereo condenser mikes with a rotary level control, switchable limiter and a mini USB connector simply to allow charging to continue while the device is in use. The mikes rotate forwards or backwards and the unit's designed for use with the iPod Touch 4th Generation, iPhone 4/4S or iPad.

The iM2 is a stereo cardioid design that claims to handle up to 125 dB input levels for high level sources such as rock concerts and motor racing, matching the recording quality of TASCAM's own DR-series sound recorders. TASCAM's own stereo PCM Recorder app is best suited to the device, offering recordings up to 12 h in length, but it's unsupported, for example, by IKM's Vocalive, which expects its audio source to come from the iPad's built-in mike or audio input socket.

The new IKM MicCast ($40) looks similar, sitting on top of an iPhone or iPad but connected to the audio socket. As its name suggests, it's designed for podcast and broadcast use, has Hi and Lo input level settings and a headphone output for direct monitoring. The pickup pattern's unidirectional, so it should be good for interviews and meetings. A desktop stand for the iPhone/iPad is included as well as VocaLive Free and iRig Recorder apps.

www.ikmultimedia.com

The Blue Microphones Mikey 2.0 at around $40 looks superficially similar too, but it's stereo and fits in the dock port. It has three sensitivity settings and unusually a minijack input, so it can be used as a stereo line input too, and a USB input, so charging can continue while it's in use. It works well with Blue's own FiRe stereo recording app but apparently there are compatibility problems with iPhone 4 and possibly with the latest iPads, so these would need to be checked out thoroughly before purchase.

MiC by Apogee, however, is another excellent iPad-oriented microphone that, like Apogee's Jam guitar input, has a proprietary multi-pin connector matched with output cables either to USB for laptop use or to a dock connector for iPhone/iPad use. Physically, the MiC falls pleasingly in size between a telecoms mike and a full-sized studio mike, so it appears robust while remaining easily transportable. It comes supplied with a small stand that resembles a budget camera tripod (an optional MiC Clip mounts it on a full-sized mike stand instead) and has a gain control on the side and input level light on the top.

Apogee MiC works well and its disadvantage over models like the Blue Yeti is simply that it lacks a headphone socket for direct zero latency monitoring.

iRig Mic from IK Multimedia ($60) takes a different approach. Connecting to the audio input, it takes the form very much of a stage vocal mike. It's a unidirectional electret condenser design and, with a minijack

FIGURE 7 Blue Mic

output on its connector, it can offer zero latency headphone monitoring. A three-level gain switch compensates for various stage and studio conditions, and of course it fits into a conventional mike stand with a suitable clamp included.

iRig Mic is accompanied by a free version of VocaLive, which is a very versatile app offering multi-effects setups of delay, chorus, pitch shift, pitch correction and much more, plus in-app purchase of additional

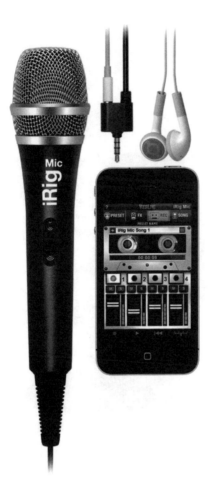

FIGURE 8   IKM iMic

effects and multitrack recording. AmpliTube Free is also included – see the sections on guitar hardware for more details – as well as the basic audio recorder iRig Free.

iRig Mic is good for stage or studio vocals, interviewing, karaoke, gaming and many more applications, very substantially improving on the performance of the built-in iPad or iPhone mike and isolating vocals from their background much more effectively – assuming you only want mono recording. It has a very substantial feel in the hand, and although not perhaps delivering the quality of Apogee's MiC that connects to the dock

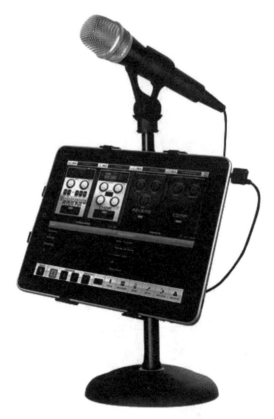

FIGURE 9   IKM Vocalive

port, its 6 ft. cable is going to make it the solution of choice for many interviewing and other applications in the field as well as in the studio.

www.ikmultimedia.com

A quick note on the Fostex AR-4i ($200) that seems a striking product mounting two mikes in an iPhone holder/dock connector (working for iPhone 4 and 4S) and offering the facility to plug in larger third-party mikes. It has an additional audio input gain control, headphone output, low-cut filter and LED meter for input level monitoring. Two AA batteries give up to 10 h of recording time and an accompanying app also called AR-4i allows some additional parameters, such as input panning and level limiting, to be set.

FIGURE 10   Fostex

The unit's really meant for increasing the quality of audio on video recordings made with the iPhone 4, a fact given away by its also including an accessory shoe onto which you may like to mount a video light.

www.fostex.com

Generic cables also exist, which can be purchased at around $30 and interface dynamic mikes such as a Shure SM58 vocal mike to the iPhone or iPad's audio socket, often offering a direct headphone minijack output for zero latency monitoring while recording – search for 'XLR Jack to iPad' on Internet sales sites.

Possibly the most flexible audio interface available for the iPad at the time of writing is the Alesis i/OMix ($350), released in mid 2012 and expanding the audio facilities of the Alesis i/ODock launched some months before by adding a four-channel input mixer. The mixer offers each of the rear panel balanced XLR/jack input channels its own control for trim, gain, pan, high and low EQ and a switchable high-pass filter, with phantom power, guitar level input switch on one channel, and quarter inch jack line and headphone outputs, while adding a switchable limiter to reduce possible recording distortion. Omitted, however, are the i/ODock's MIDI input and output, so this is a device more for recordists than for MIDI musicians.

FIGURE 11 Alesis ioMIX

Newly added is the ability to mount on a mike stand using the optional Alesis Module Mount and a hinged cover that completely covers the mounted iPad for safety. An adapter tray ensures original model or iPad 2 or new iPad models all fit, and the composite video RCA phono output is now accompanied by an S-video output. The iPad 2's HDMI video output ability isn't supported, but for video projection purposes, the composite and S-video facilities are more likely to be relevant.

Slipping an iPad inside an i/OMix affords the user a fairly hefty piece of equipment, certainly comparable in size to carry around a laptop plus a small multi-channel USB or Firewire audio/MIDI interface. Studio software on a laptop will certainly be more powerful than the iPad plus i/OMix combination (just contrast the iPad version and the much more powerful laptop version of GarageBand, for example, or consider the iPad's lack of anything comparable to Cubase or Logic

Express) and so the i/OMix might be best considered as a purchase for those already heavily committed to use only an iPad as a recording medium and wishing to improve its facilities in group recording, karaoke, audio-visual presentation and other areas of work. With separate Main Out and Headphone Out volume controls, the i/OMix should be versatile enough to give the user the flexibility to work in all sorts of studio, stage and presentation settings, though it would depend on the app in use whether you could – for example – play a backing video with sound while mixing in a synthesizer, guitar and microphone from the rear connections.

www.alesis.com

Incidentally, the iPad is starting to find use in many mixing and concert sound applications, either as a remote controller for a sound system or increasingly as the processing element of a lighting or sound desk. We won't look into these in too much detail since these products really use an iPad as an add-on, rather than acting as an add-on to an iPad. However, this is bound to become an increasing trend as manufacturers see the potential in having most of the processing work carried out by an easily updateable independent element of their design.

FIGURE 12 Behringer

One example of this trend is the XENYX iX series of three live performance mixers from Behringer, which take advantage of the iPad's video out facility to sync performances to video. The mixers include USB out to record audio to a computer and built-in Klark-Teknik audio effects, all controlled via an app on the iPad.

www.behringer.com

Mackie's new DL1608 mixer goes even further in that the iPad does almost all of the work and provides the control surfaces of the mixer – though the mixer alone costs more than $1000, so it must include a lot of input/output technology. In fact, it features 16 *Onyx mic preamps* and 24 bit AD/DA converters, and the associated iPad can be used either mounted inside or remote from the mixer to control the mix as well as plug-ins like EQ, dynamics and effects.

FIGURE 13 Mackie

www.mackie.com

Much simpler, but not yet released at the time of writing, is the iConnectMUSE, a small mixing box offering six stereo inputs and six stereo outputs each with its own unique mix, plus a headphone output, all run from an iPad. The inputs include quarter inch analog, USB from a computer and Ethernet connectors, so the $250 box might be useful for mixing together all sorts of media sources.

FIGURE 14   iConnect

www.iconnectivity.com

Routing audio into the iPad is bound to become a more and more common practise, and when apps are developed which handle multi-track recording more readily, multi-channel input interfaces are bound to appear. As processing speeds become higher with the release of successive iPad models, its use in audio applications will become more and more widespread.

Chapter 4

MIDI Interfaces

MIDI interfacing for controlling notes, patch changes and other parameters is still commonly found on musical instruments, and the serial communications protocol used by MIDI is readily handled by the iPad.

The Line 6 MIDI Mobilizer was the first MIDI interface for the iPad to hit the market and is currently on Version II ($70) which supports Core MIDI (any old Model 1 units still available should be avoided). MIDI Mobilizer II remains an obvious choice for anyone who wants to add MIDI to an iPad.

The interface itself is tiny, barely larger than an Apple Camera Connection Kit adapter, although finished in black and featuring nothing more than a dock connector on the front and MIDI In/Out stereo mini-jack sockets on the rear. Minijack was, of course, not the original socket specified for the MIDI protocol – the five-pin DIN socket has always been the standard, despite the fact that only three pins were in any way active – and so stereo minijack to five-pin DIN adapter cables are supplied. A disadvantage, of course, is that your iPad can't remain on charge while using the MIDI Mobilizer.

MIDI Mobilizer II works faultlessly with Core MIDI apps and does handle System Exclusive, so as well as controlling note, velocity, pitch bend, modulation and other functions can be used to dump and restore patches and carry out sound editing, provided you have the apps to do so. Line 6 provides MIDI Memo, a free data recorder app for just this

FIGURE 1   Line 6 MIDI

purpose, and they really have in mind dumping patches from their guitar effects units such as the Pod, or from synthesizers such as the Alesis
Q series. Sys Ex recordings simply appear as a list of Memos in the app
and can be replayed or sent by e-mail at any time – these could include
lighting cues for programmable systems, as well as SMF (Standard MIDI
File) songs and all sorts of other musical data. MIDI Memo Recorder
also handles any required firmware updates for the MIDI Mobilizer.

Line 6 publishes a long list of MIDI Mobilizer-compatible apps, including the Luminair DMX lighting controller, Wavesynth, Fairlight sampler
app, Sunrizer synth and many more, which you can see in full here –

www.line6.com/midimobilizer/mm/more_apps.html

IK Multimedia's iRig MIDI ($70) looks superficially similar and again
uses stereo minijack sockets rather than five-pin DIN sockets, but here

there's three of them for MIDI In, Out and Thru. A micro USB port allows your iPhone or iPad to remain on charge while the device is in use. With the dock connector inserted you just load up the free IKM iRig MIDI Recorder app and all Core MIDI apps should work instantly. The MIDI Recorder app takes System Exclusive and other types of data dump so can be used to archive patches and sound files, which can be sent by e-mail, iTunes File Sharing or WiFi.

FIGURE 2   IKM iRig MIDI

One good application for iRig MIDI is to load up IKM's SampleTank app which offers many high-quality keyboard, percussion and other instrumental and sound effects samples. Coupled with a good quality MIDI controller keyboard, you're looking at a stage or studio perform-ance system to compare favourably with keyboard instruments and modules costing thousands of dollars just a few years ago.

www.ikmultimedia.com

YAMAHA's i-MX1 ($100 or considerably less) is another MIDI interface which looks very similar to the Line 6 product – it's a tiny black cartridge connected to the dock port. It works with iPhone or iPad on Core MIDI applications, but there are also dedicated applications available for Yamaha synthesizers which will expand the functions of the instrument. The 'Faders & XY Pad MIDI Control' app controls a synthesizer's sounds by sending MIDI control change messages from the iPad, whereas 'Keyboard Arp & Drum Pad' is an application which transmits the notes to the synthesizer.

FIGURE 3 Yamaha MIDI

www.yamaha.com

Just appearing at the time of writing is the Vestax VMIDI which connects USB MIDI devices to iOS devices. The circular unit includes a short captive cable to a dock connector, and the VMIDI keeps your iOS device on charge while in use. All core MIDI apps should be supported (Vestax claim over 100 are in existence) and the VMIDI also has a boosted audio output with a large volume control. The MIDI connection from USB is a mini-DIN so an adapter cable is included, and Vestax see the interface routing keyboard controllers or any of their line of DJ controllers into an iOS device, with an output to headphones and (using a splitter) to speakers or other amplification.

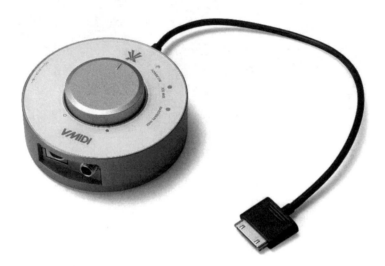

FIGURE 4   Vestax MIDI

www.vestax.com

The ultimate solution for MIDI and iOS interfacing may be the iConnectMIDI from iConnectivity, which interfaces five-pin MIDI, USB MIDI, Mac, PC and iOS devices all at the same time. The unit in a small metal box handles MIDI, up to eight USB sources using standard hubs, and two iOS or computer ports simultaneously, so is likely to prove the go-to solution for more complex setups involving iOS devices. Here's one of the manufacturer's examples of what advanced tasks the device could carry out – 'Want your iPad to talk to just one keyboard or all connected instruments? No problem. Or want to filter out all pitch bend

FIGURE 5   iConnectivity

data going out of ports 3 and 7? Done. Have all the data from three keyboard controllers go to one connected synthesizer? Easy'.

www.iconnectivity.com

MIDI interfacing on the iPad allows the device to act as a powerful sound module slaved to a keyboard or sequencer, as a sequencer playing external modules, or as a drum machine providing clock signals to synchronise other instruments. For those interested in using both MIDI interfacing and audio, combined interfaces are covered in Chapter 5.

Chapter 5

Audio/MIDI Interfaces

Just a few iPad interfaces now combine MIDI and audio abilities, together with other facilities such as video interfacing.

The first of these to be released at around $300 was the Alesis i/ODock, an ambitious product which aimed to turn the original model iPad, when accompanied by suitable apps, into a fully fledged recording studio. The i/ODock comprises a wedge-shaped support stand, with an external power supply, into which the iPad slots, offering on its sides and rear MIDI In and Out sockets (of the familiar five-pin DIN variety), a USB socket (for connection to a computer only, not to a USB keyboard controller), balanced XLR/jack microphone sockets with volume controls and switchable phantom power, each switchable to line and one to guitar level, stereo audio quarter-inch jack outputs with volume control, a headphone output with volume control, a footswitch socket, and a video output which on the original iPad would only send an image when playing the Movie app or one or two others. On the iPad 2 and new iPad, a small adapter tray compensates for the thinner body size, and the video output, of course, now shows screen images of all apps in use.

This is a formidable specification, and the selling price falls far below that of acquiring separate MIDI and audio interfaces, phantom power supply, video output cable and all the appropriate socket convertors. An iPad mounted in an i/ODock does indeed become a formidable looking piece of studio equipment. Both Alesis and Gator offer a laptop-sized backpack for transporting an iPad inside an i/ODock at around $50.

FIGURE 1   Alesis IODock

The i/ODock, however, has received mixed reviews, although in cases where it doesn't perform well, this is more likely to be the fault of the specific app rather than of the hardware, and this in turn is often likely to be caused by Apple making iOS updates without any particular regard to how existing apps have been designed. The MIDI In/Out facilities of the i/ODock work extremely well for many synthesizer apps when combined with most MIDI controller keyboards. Whether functions like the footswitch input (most usually used for Sustain) or filter cut-off will work predictably depends more on the individual app. Again, the balanced XLR audio inputs work extremely well when combined with apps which expect this type of input appearing via the Dock connector – those which expect an input through the iPad's Audio In would fare less well.

At various points in time, the i/ODock has failed completely to perform with some apps. It is possible to download i/ODock firmware updates from Alesis, but these aren't always able to keep up with iOS updates from Apple. iOS5 apparently worked extremely poorly with the i/ODock, while iOS 5.1 solved most of the problems. No version of firmware and iOS to date has allowed the i/ODock's MIDI sockets to handle System Exclusive messages, which is very odd, and makes it of limited use for various types of synthesizer patch librarian and controller packages.

The construction quality of the i/ODock is fair – it's lightweight yet reasonably solid, perhaps not designed with regular onstage performance in mind – and its requirement for a chunky external power supply is more or less dictated by its ability to offer phantom power to XLR mikes.

Offering slightly fewer facilities than the i/ODock but perhaps avoiding some of its problems and released around mid-2012, the TASCAM iU2 at around $200 comprises a USB or dock-powered audio and MIDI interface offering phantom power for one or two microphones. On its top panel are rotary controls for Input L and R levels, monitor mix (between the input source and a host computer), Line Out Level and Headphones Out Level. On the front panel are the quarter-inch stereo-balanced audio inputs (adapter cables to female XLRs are supplied) and a mono quarter-inch guitar input. On the rear panel are stereo RCA phono line audio outputs, an RCA digital stereo output (not commonly used these days, but presumably still found on some TASCAM recording products), mini USB output (which can be used to charge iOS devices with a suitable cable, if it's not being used to interface to a laptop), plus a captive dock cable, whereas on the left panel are a stereo minijack headphone output and two MIDI connections – as on the IK Multimedia product, offered as stereo minijacks with conversion cables to five-pin DIN plugs included. Switches under the unit select iOS or USB hosts, phantom power on/off, and mike/line level for the inputs (with guitar level as an option for the Left input).

Although compact – about half the footfall of an iPad but twice as thick – the iU2 offers a lot of possibilities for recording on a laptop or on the iPod Touch, iPhone or iPad. Direct monitoring of the input source can overcome any latency problems, and the source can be anything from a cassette or CD player or other line level source to dynamic mikes, phantom-powered condenser mikes or a guitar. Note that phantom-powering mikes while using an iOS host such as an iPad requires a USB charger to be connected to the mini USB socket.

FIGURE 2   TASCAM iU2

TASCAM offers the free stereo PCM Recorder app and their four-channel Portastudio recording app for iOS, but other apps work equally well with the iU2. The unit's MIDI implementation is comprehensive, with all sorts of signals including velocity, channel and polyphonic after-touch and MIDI clocks being supported, although which of these is recognised by your app will vary. Tested with a Korg NanoKey 2, the iU2 performed well supplying MIDI to Sunrizer and other synth apps as well as audio to GarageBand.

Just coming onto the market at the time of writing and somewhere in design philosophy between the Alesis i/ODock and the TASCAM iU2 stands the Griffin StudioConnect ($150), which is a wedge-shaped stand mounting an iPad vertically and offering a headphone output on the front together with a volume rotary and large data wheel style controller, and on the rear, five-pin DIN MIDI In and Out sockets, RCA phono stereo audio outputs, stereo minijack audio in and a quarter-inch jack guitar input. The device has a short captive cable to the iPad dock port and keeps the iPad on charge while in use.

FIGURE 3 Griffin StudioConnect

StudioConnect is designed particularly to work with GarageBand and while not offering the wide range of connections of the Alesis i/ODock such as video out and balanced mike in, does offer a compact method of bringing the iPad into use as a mobile studio – as long as you don't want to record a stereo synth. However, since it features MIDI rather than USB ports, and no phantom-powered XLR mike port, almost everything you connect to it will need its own power supply as well, so it may not be entirely the portable solution it seems.

www.griffintechnology.com

No doubt the number of all-in-one, MIDI and audio interfaces for the iPad will increase in the future. It's also likely, given the extended video shooting abilities of the latest iPad, that these will offer a lot more in the way of video outputs as well as audio and MIDI interfacing.

Chapter 6

Keyboards, DJ Decks and Controllers

Since USB-equipped products can be interfaced to the iPad using the Apple Camera Connection Kit or its clones and MIDI-equipped products can be interfaced using any one of the number of available MIDI interfaces, the number of keyboard and controller products which can potentially be used with the iPad is pretty much unlimited. It's best to determine a suitable product according to your exact studio or stage application.

If you want to play complex piano parts, perhaps taking advantage of the high-quality piano sounds within Garageband, then a full-length weighted keyboard equipped with either USB or MIDI would be ideal, perhaps a model from Akai, Yamaha, M-Audio or Peavey. These usually won't be powered from the iPad and will have their own mains power cable, external power unit or possibly USB power input from a computer. The advantage of a large keyboard is that an iPad could sit on it like a sheet music stand. One excellent model is the M-Audio KeyStudio88 which offers an enormous number of controller slider and rotaries and excellent weighted keyboard action at a ridiculously low price. It's on the bulky side, but the company can also offer a more compact ProKeysSono88 model which lacks most of the controllers but offers internal piano and other sounds.

For playing synthesizer or sampled parts onstage (perhaps sourced from IK Multimedia's Sampletank app) then a smaller plastic-key USB or MIDI controller would be ideal, and there's a vast range of models available from Roland, Korg, Behringer, M-Audio, Novation, E-Mu, Alesis and many other manufacturers.

For moving around on stage, there are many smaller keyboard controllers which can be picked up easily, or guitar-style strap-on keyboard controllers including the new Alesis Vortex to consider. For drummers, there are several MIDI or USB-equipped pad sets, plus Akai's MPC Fly which is a hardware/app combination for the iPad simulating the company's MPC sampling drum machine line.

Very few of these products are specifically dedicated to the iPad, aimed as they are at the wider market of computer and MIDI musicians – so we won't be attempting to list them in any comprehensive manner, but will note any which may be particularly useful alongside specific iPad apps, together with the very few models which are now appearing with dedicated interfacing to the iPad dock connector.

Controller keyboards will typically offer velocity sensitive keys, a pitch bend and modulation device (except in the case of very small

FIGURE 1 Mark Jenkins Alien

models like the Evolution eKeys which I've used on stage through the CCK), octave switching particularly in the case of a short keyboard, usually some kind of data controller or maybe even four or eight or more, and occasionally drum pads or maybe an XY pad. Whether these facilities work with a specific iPad app will depend on the app. If apps are programmed using Core MIDI, they should at least respond to the basics, but a piano app, for example, may not respond to pitch bend, and a synthesizer app may or may not offer control of filter cutoff from a data slider (GarageBand doesn't, Sunrizer can). Happily, the iPad is small enough to take into any store along with a Camera Connection Kit to try out its response to a particular controller keyboard using any favourite musical app, before making a purchasing decision.

FIGURE 2 Novation

One range definitely well suited to use with the iPad – compact, affordable and well specified – is the USB-equipped Microkeys range from Korg. Unfortunately, the original 37-key model requires too much power to run via the Camera Connection Kit, so a powered USB inter-face must be used – but smaller 37- and larger 61-key models are also available now.

One range which does work directly through the Camera Connec-tion Kit though is the Korg Nano range including the NanoKey 2 ($50, in black or white), which despite its button-like two octaves of keys, plays surprisingly well, offering velocity response, Octave Up and Down buttons, a Modulation button and Pitch Bend Up/Down buttons. The NanoKey, barely wider than an iPad, makes an excellent portable key-board solution and is also great for playing drum sounds or triggering samples, if an alternative to the on-screen virtual keyboard of any particular app is required.

www.korg.com

M-Audio's Keystation Mini 32 ($100) is also extremely compact, offering 2½ octaves of mini keys. Velocity response curves can be programmed, and there's one specific for drum programming. There are four assignable controls including a rotary knob and Mac/PC music composition software are included, but you will have to add a Camera Connection Kit to use the keyboard.

FIGURE 3 M-Audio

www.m-audio.com

One line of controller keyboards which is at least in part specifically dedicated to the iPad is the Mobile Keys from Line 6, in velocity sensitive 25 note ($150) and 49 note ($200) versions. These are the world's first triple output controller keyboards featuring MIDI, USB and an output to the dock port for iOS devices. Other than the number of keys they're identical, featuring pitch bend and modulation wheels, volume and pan rotaries, Octave Up and Down buttons, sustain and expression foot controller inputs.

The Line 6 models play well and are ideally compatible with Garageband and apps like Pianist Pro. Their only disadvantage is the non-standard nature of the supplied iOS cable, so make sure you don't lose it.

Samson's Carbon 49 ($90) is a similar four-octave keyboard with USB and MIDI only, so you'll need a Camera Connection Kit to use it with an iPad, but it does have a slot to mount an iPad like a sheet music stand on the top panel. It has transpose and octave buttons and a sustain

FIGURE 4   Line 6

FIGURE 5   Samson Carbon

footswitch input along with pitch bend and modulation wheels, and a Shift key for adjusting more than a dozen performance-related parameters. Included is the Native Instruments' Komplete Elements software for PC and Mac, and users have found that the Carbon 49 can be powered via the CCK. Even more sophisticated is the Samson Graphite

FIGURE 6 Samson Graphite

model, which can also be powered from an iPad and boasts a great looking battery of controllers and drum pads, but it doesn't specifically mount the iPad on its top panel.

www.samson.com

In fact, there are quite a few keyboard designs which do physically accommodate an iOS device rather than just interfacing to it. The PocketLoops from Gear4.com is a two-octave keyboard which just accommodates an iPhone, running an app which helps create house, hip-hop, reggae and other styles of loops, offering 13 synth, piano and other sounds, but the Piano Apprentice from Ion accommodates an iPad in the manner of a music rest, and runs piano tuition apps which light the two octaves of keyboard keys as they play.

FIGURE 7 Ion Piano

More closely dedicated to iPad use is the Akai Synthstation 49, which developed from the Synthstation 25 for iPhone/iPod Touch.

Synthstation 25 ($75) offers two octaves of mini keys, pitch bend and modulation wheels and mounts an iPhone horizontally on its top panel. The Synthstation 49 ($250) looks like a much more substantial piece of kit, offering four octaves of full-sized velocity sensitive keys, various controllers and a set of nine drum pads resembling one of Akai's MPC line sampling drum machines.

There are transport controls including Play, Pause and Record, as well as USB and MIDI connections, quarter-inch stereo audio outputs and a headphone output, and the Synthstation charges your iPad while it's connected.

Both models work well with Akai's SynthStation app which offers three polyphonic analogue-style synths, arpeggiator, drums with multiple kits, touch pads and variable pitch, touch effects XY pad and song

FIGURE 8   Akai 25

FIGURE 9   Akai 49

composition – in fact the controller layout on the keyboard more or less mirrors the facilities of the app, and whether other apps work with equal ease will depend on how they're implemented. The developer's kit at www.synthstationapp.com allows third parties to write for the Synthstation 49, although all existing Core MIDI apps should already work with the keyboard to some extent.

About to launch at the time of writing, the Alesis Vortex ($300) harks back to the days of strap-on controller keyboards like the Moog Liberation, the MIDI-equipped Yamaha SHS10, SH200, KX5 and KX1, the Casio AZ1 and various Roland models right up to the currently available AX Synth. Guitar-style keyboards have had a bad press in some areas but are great fun to use if their controller layout is sympathetic to your playing style. Past users of strap-on keyboards include Jean-Michel Jarre (Moog Liberation, Yamaha KX5 and custom controllers from Lag), Rick Wakeman (Roland AX1), George Duke (Clavitar), Jan Hammer (who started with a custom separated MiniMoog keyboard before using Lync and other controllers) and many others.

The Alesis Vortex is a relatively compact model with three octaves of full-sized keys and offers both MIDI and USB with onboard battery power. It has a modulation wheel and a pitch bend strip, and like the iPad includes an accelerometer so spatial movements can send controller signals and create effects like volume swells, pitch bend or wah-wah.

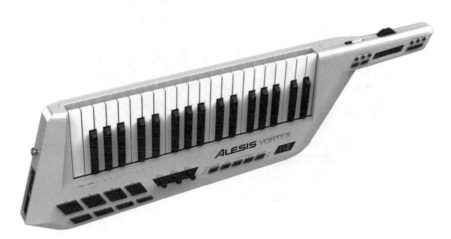

FIGURE 10 Alesis Vortex

Unusually, the keyboard also features transport controls so you can start and stop sequences remotely, as well as eight drum pads, and internal battery power means it can run with iOS devices using the Camera Connection Kit or other interface.

At the time of writing, there's surprisingly little on the market in the way of iPad controllers for drummers, unless you count the pads built into many of the USB keyboards available. The iPad boasts loads of drum apps range from very simple free examples offering a few sounds, to highly programmable sample playback software such as SampleTank. First out of the trap in the way of sophisticated drum interfacing is likely to be the DM Dock from Alesis ($300), a mate to the company's i/ODock and AMPDock models specifically aimed at playing the drum sounds of GarageBand and other apps from a set of trigger pads. Alesis, of course, manufacture a wide range of pads to accompany their DMX10 and other hardware electronic kits, and you can interface up to 13 pads or other types of trigger input through the DM Dock.

FIGURE 11 Alesis DM Dock

The DM Dock is finished in silver and the front panel just features main and headphone volume controls. The rear panel features the quarter-inch jack audio outputs as well as MIDI In/Out and USB sockets, alongside 14 TRS quarter-inch jack trigger inputs and a HiHat control input, and there's another footswitch input on the side to select a new kit, tap in a tempo or start and stop sequences, as well as a mini-jack audio input to mix in, for example, an MP3 player. The DM Dock charges an iPad while playing, and the optional Alesis Module Mount supports the device on a drum stand or rack. It looks as if an Alesis app called DMTouch will accompany the device and allow large numbers of different drum sounds to be loaded and saved as kits, just as happens on the Alesis hardware drum modules such as the DM5.

The DM Dock is compatible with single- or dual-zone drum and cymbal pads, triple-zone ride cymbals, continuous control hi-hats, multiple-choke cymbals and acoustic drum triggers from various manufacturers. With luck, the inputs will also be sensitive enough to trigger from piezo pickups which are a few pennies from any electronics store. This gives the potential of triggering iPad sounds remotely from any surface such as pieces of furniture, sculptures, articles of clothing and so on, which is an exciting possibility.

www.alesis.com

Another drum-oriented device is the MPC Fly from Akai. This takes the form of a small iPad sleeve and an accompanying app which simulates the company's MPC line of sampling drum machines. The front of the sleeve, rather than featuring a Bluetooth keyboard as found in many inexpensive designs, includes 16 physical pads as found on the MPC machines themselves, and so the iPad becomes a tiny portable MPC including a four-track mixer preloaded with samples and drums. The app allows for sampling and composing with sounds using the iPad's mike or line in, or from the iTunes library, and finished compositions can be shared on Facebook, Twitter and SoundCloud or exported to a computer. Price still to be confirmed at the time of writing.

www.akai.com

FIGURE 12   Akai MPC Fly

No doubt many more percussion controllers for the iPad will appear as time goes on, and it will be up to the app designers to make sure that triggering from pads is as accessible and reliable as triggering by hand from the iPad screen itself.

DJ CONTROLLERS

The iPad has become a powerful resource for DJs, and there are many excellent DJ-oriented apps available. Some of these, like GrooveMaker from IK Multimedia and RJ Voyager, offer to create new sets of loops and beats, but most are intended just to play back existing music files in a familiar two-deck-plus-mixer format. Some of these DJ apps work incredibly well and a personal favourite is Air Scratch HD, which offers a beautiful simulation of two vinyl decks with easy access to tracks through iTunes, volume meters, varispeed, tap tempo, beat matching, looping, equalisers and filter effects.

Adding a physical DJ-style controller to the iPad when its existing touch-responsive screen display might appear pretty much up to the ask could appear counterproductive, but many hardware DJ controllers are now appearing and most of them will certainly give improved access to the various facilities of DJ apps.

IK Multimedia offers the DJ Rig app which again is a dual vinyl deck simulation, running on either iPhone or iPad. Just launching at the time of writing is the company's iRig MIX ($100), a compact hardware mixer with two audio inputs, crossfader, bass and treble EQ, cue switch on each channel to preview in headphones, and the ability to work with either one or two iOS device sources (with a single device, its sound will be split into dual mono so advance cueing is still possible), or indeed with MP3 players or other sound sources. When used with the DJ Rig app, automatic tempo matching and beat syncing are available.

iRig MIX is also aimed at solo musicians other than DJ's, since it offers an extra guitar/mike input maybe for guitar processed through the company's AmpliTube app, or a voice processed through their VocaLive package. Audio outputs are RCA phono stereo line outputs, so the iRig MIX – the same length as an iPad, or about twice

FIGURE 13   IKM iRig MIX

the size of an iPhone – offers a compact way to interface to any conventional sound system. Power is from batteries, mains transformer
or USB.

Numark's iDJ (sometimes know as the iOS DJ, $90) is a very afford-
able hardware DJ controller which mounts an iPad (or an iPhone or
iPod) vertically about its dual simulated vinyl decks. Algoriddim's djay
is the recommended app, although others will work too, and the hard-
ware/app combination simply accesses your iTunes library offering
scratching, looping, crossfade and EQ abilities.

The hardware looks a little toy-like and a major criticism is that
it's powered from the iPad, draining its charge quickly. However,
this could be a very affordable entry into iPad-based DJ work. Ion's
IDJ2Go ($70) is smaller but looks more substantial, again intended

FIGURE 14 Numark IDJ

for Algoriddim djay on the iPad (though it has an app of its own too) but physically not much wider than the iPad itself and needing no external power.

FIGURE 15 Numark IDJ Pro

Numark, of course, has some much more upmarket models such as the $400 iDJ3 which mounts a single iPhone or iPod, various USB or MIDI-equipped controllers, and decks which mix one or two iPhones, but a new pro quality iPad-oriented model is just coming on the market. The iDJPro ($500) mounts an iPad flush with its aluminium top panel casing, and although still oriented towards the use of the same djay app, it offers much higher quality controls and extended functions. Your music library is now scrolled through with a dedicated knob and the unit's compatible with AirPlay, so you can play out to wireless speakers. Balanced XLR master outputs mean that connection to much more professional sound systems can easily be made.

www.numark.com

Several other companies offer DJ-oriented MIDI controllers, Vestax, for example, having many models in their VC, TR and Typhoon lines. Most of these have two deck controllers and all the usual mixing

FIGURE 16   Vestax VCI

facilities, the smaller Pad-One offering just drum pads and an XY pad controller, the VCM models just offering mixing without deck control. All these are USB-equipped controllers intended for use with DJ software such as Serato, but using a MIDI interface – such as the VMIDI from Vestax mentioned earlier in Chapter 5.

www.vestax.com

Another company with a very wide range of MIDI controllers including keyboards, DJ and drum pad controllers, guitar and mike inputs for iPad (iPlugG and iPlug M) is Icon, based in Italy. The company offers two-, three- and five-octave keyboards, small pad controllers similar to the Korg Nano range, and an unusual multi-purpose light matrix controller called iCreative which can play notes, arpeggios or drum events. It's not specifically iPad compatible but like most MIDI controllers should have something to offer in conjuction with any Core MIDI apps.

www.icon-global.com

The iPad is undoubtedly becoming a more and more serious resource for DJ's and VJ's, and many more DJ-oriented deck controllers can be expected to appear in the near future.

A list of keyboards, controllers, synthesizers, audio inputs and other devices thought to be compatible with iOS devices, together with some discussion of the latest developments can be found here:

http://iosmidi.com

The range of keyboard and other controllers which can potentially be used with the iPad is extremely large, but some apps designers need to put more effort into ensuring that controller movements are responded to in some musically useful manner. In the world of MIDI music, there are many unusual types of controllers available, including pitch ribbons and pressure sensitive pads, but since the touch screen of the iPad itself is a superb control surface which works well with theremin-type apps like MorphWiz, there's perhaps less incentive than there might be to develop unusual physical controllers for the iPad.

Chapter 7

For the Guitarist

We looked in Chapters 3 and 5 at some input devices aimed either at general purpose audio use or at the guitarist specifically, and the iPhone and iPad have developed a very good record for hosting guitar multi-effects, rehearsal and recording apps, with the guitar input derived either from the audio socket or from the dock connector. This has led to a busy market in specialist guitar input devices, ranging from a simple cable with the appropriate line level buffering of the type already discussed to complex pedalboards usually matched with very highly programmable guitar multi-effects apps.

It must be said when considering the use of an iPhone or iPad as a guitar effects processor that some extremely good and compact hardware effects like those from Zoom, or the Pandora range from Korg, have provided a sturdy and reliable solution for years. Perhaps it isn't worth tying up a $900 iPad to do the job of a $90 guitar pedal, although the iPad does certainly offer some attractive screen displays including great highly visible tuning displays, and some highly variable and programmable guitar-oriented apps.

Before starting to use an iOS device in guitar applications, you're going to need to decide where to place it for the most effective use in the studio or onstage. Guitar Sidekick at $25 is a handy grip for mounting a tuner, or equally an iPhone running a guitar effects app, onto your guitar head – although an iPad doing the same job would be too large and much better placed on a mike stand, using IKM's iKlip or another type of support bracket.

Peavey AmpKit Link ($40) is one of the many devices offering a jack input from your guitar, a minijack output to the iPad and a minijack headphone socket for monitoring. It runs from batteries and so claims to avoid the crosstalk problems inherent with unpowered input/output devices, as well as compensating for the iPad audio input's low-frequency roll-off oriented towards use on the human voice.

FIGURE 1   Peavey

The free AmpKit app by Agile Partners offers three pedal simulations, two cabinet and two mike simulations, with many more available through in-app purchase, but AmpKit Link works with any app (for example, GuitarToolKit) and can be used simply as an audio input without any effects added. Adding all the effects you want to AmpKit can add $25 to $100 to the purchase price, and it must be said that while there are many satisfied users of AmpKit Link, some have complained of unreliability and poor socket construction.

More recently released is AmpKitLinkHD (no price at the time of writing), which uses the dock connector and so can promise better crosstalk and noise performance. An optional power adapter means the iPad doesn't lose the ability to stay on charge while it's in use. Again, the device offers a guitar input, headphone and line output, and since a USB cable is also included, the device can work for Mac and PC recording as well as iOS setups.

www.peavey.com

www.agilepartners.com

Griffin's Stompbox ($100) looks deceptively simple – just four footswitches in a rounded-corner box, with a quarter-inch input for your guitar. It works only with iShred Live, a free app which comes with an amp simulator, delay, flanger and noise gate as standard, with phaser, treble booster, chorus, auto wah, a recorder and other effects as in-app purchases. iShred Live works with other input devices, but the four switches on Stompbox are optimised to work with the app – holding a switch down, for example, changes function from bypassing effects to choosing a new preset and to starting up a metronome. A quarter-inch socket adds an optional wah or volume pedal and included with the package is the Griffin GuitarConnect cable (usually $30) which connects the guitar to the iPad. This cable is 6-feet long and simply offers a quarter-inch jack plug for the guitar, a minijack headphone socket next to it for direct monitoring and a minijack plug into the iPad.

www.griffintechnology.com

The sound of iShredLive is as versatile as that from AmpliTube and other multi-effects apps, but some players will prefer alternative multi-effects apps with which the Stompbox isn't compatible.

At around $300, the Alesis AMPDock modifies the layout of the i/ODock with the needs of the guitarist in mind. Inputs are balanced XLR, a quarter-inch jack line or guitar, with quarter-inch outputs to a mixing desk or PA system, plus MIDI and headphone connections. The AMPDock comes complete with a well-constructed pedal board which can control volume, wah or other parameters and change patches or bypass the system, or using MIDI In, you can interface pedal boards and controllers from other manufacturers.

FIGURE 2 Alesis AMPDOCK

Recommended apps for AMPDock include GarageBand, AmpliTube and JamUp. Four-front panel controls are programmable but will generally access Main, Headphone, Input 1 and Input 2 levels, whereas two assignable knobs can control various parameters like drive level or delay time. The AMPDock includes a USB port so can also be used with audio/MIDI applications on a computer.

A small kick stand allows the AMPDock to sit over the carrying handle found on the top of most guitar amps. It will also mount on a mike stand using the optional Alesis Module Mount, if you prefer not to have your valuable iPad anywhere near other items being stomped on at floor level.

www.alesis.com

Digitech's IPB10 ($600) is similar to the Alesis AMPDock but is designed largely for floor mounting – although some users are already adding a dock extension cable so their iPad doesn't actually have to remain on the floor inside it. The accompanying app offers a drag and drop design to arrange 87 pedals, 54 amps and 26 cabinet simulations, and 100 presets are footswitch accessible, although there's no limit on the total number of presets which can be saved.

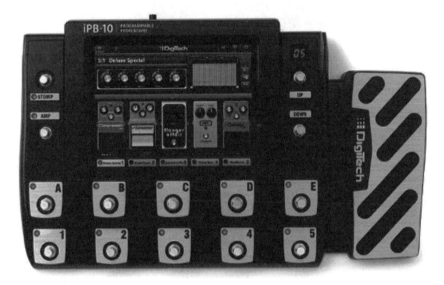

FIGURE 3 Digitech IPB

The IPB10 has 10 footswitches and a swell pedal built in, but for a much simpler piece of hardware, Digitech offer iStomp, a small single stomp box with four controls, footswitch, stereo input and output, which can be reprogrammed from the Stomp Shop app to act as any one of 32 different pedals (with more to come, no doubt). An overdrive and a delay model is included, other pedal simulations cost from a dollar upwards and you can try each one for 5 min before buying. A special cable is needed to download from an iOS device to the pedal.

www.digitech.com

FIGURE 4   Digitech iStomp

Chapter 8

Stands, Cases and Accessories

The iPad has generated a huge third-party market in stands, cases, covers, keyboards, speakers, docks, chargers and accessory cables, some more relevant for musical applications than others. It's important to realise that although some desktop iPad stands can retail for $30, $60 or even $100, the iPad market is large enough so that dollar stores (pound shops) can usually afford to offer something perfectly adequate for supporting and viewing an iPad in either landscape or portrait orientation.

Hard rear shells, screen covers, extendable sync/charging cables, audio output splitters and headphone to RCA adapter cables are also usually available at a similar bargain price. Very basic slip cases for the iPad are also available inexpensively, though if the device has become your main musical resource on the road as well as in the studio, you may wish to spend a little more on protecting it. The rubberised Griffin Survivor is a good option, but there's also a wide choice of covers from Cygnett including the Aerosphere ($25). This is a very attractive bubble-covered sleeve in orange, black, brown and other colours with internal elastic fasteners, which gives a sensation of great protection for an iPad without adding too much weight or bulk.

us.cygnett.com

Simple charger and audio splitter cables for the iPad are easily available, but specialised cables will offer a little more. Griffin markets the

GC20007 DJ Cable ($20) designed for the dual-deck 'djay' app by Algoriddm. The cable splits the audio output, and with Enable Split Output selected in the app, you'll hear the master output (the currently playing song) from the sound system and the cue output (the next song ready to be played) in headphones. Both sound outputs are binaural, in other words, mono on both channels. Enabling advance cueing is really vital for DJ purposes, though this cable will limit your onstage sound to dual mono rather than stereo. Griffin also markets a vast range of stands, cases, chargers and other accessories for the iPad, iPhone and other tablets and smartphones, their Survivor case being a good option for iPads on the road – and also see sections on their guitar and audio/ MIDI interface products.

An important aspect when you start using an iPad under stage or studio conditions is accessibility and being able to mount the device where it's readily useable. Other than desktop stands, the iKlip from IK Multimedia was one of the first available more professional solutions to this question, mounting the original iPad and later the iPad 2 in either landscape or portrait orientation onto any conventional microphone stand.

FIGURE 1 iKlip

Most musicians will have access to mike stands at any venue, and the iKlip's mounting sleeve is wide enough to slip over any standard design. Tighten the rear screw, and the iPad is firmly mounted exactly where required. Changing from landscape to portrait orientation requires a screwdriver which is a small limitation, but in my own performance setup using two iKlips that didn't vary. During performance, I wanted two iPads to face the audience, not simply look up at me from a desk, so would mount them slightly behind my performance position using iKlips and whatever mike stands available at the venue, typically running Beep Street iSequence on one side and Garageband plus Sunrizer for instrument sounds, such as rock organ or synthesizer, on the other side.

FIGURE 2 M Jenkins USA Concert

The iKlips proved to be an excellent resource and are highly recommended – though once you've mounted the iPads in the air, there's a great deal of thought to be put into cabling them up securely, particularly if small and delicate Camera Connection Kits or small MIDI interfaces are involved. An alternative is the simpler iKlip Studio, which is a desktop stand. In addition, an iKlip MINI version is now available for iPhones and this comes complete with a bracket for mounting the iRig guitar input, since there'll be a lot of use of the iPhone as an effects unit with IKM's own AmpliTube guitar effects app.

FIGURE 3 Millenium

Now, there are other iPad 'music stand' designs, including one in German mail order specialist Thomann's own-product line Millenium. The Millenium Tablet Holder is a universal tablet holder for the iPad, Samsung Galaxy and so on, and offers a plastic plate with adjustable clamps suitable for any device up to 28 cm × 21 cm. A versatile clamp holder allows it to be mounted with equal ease onto desks or

FIGURE 4 K and M

microphone stands. Thomann also stocked the K&M line of tablet hold-ers, including the 19722 iPad Holder for iPad 2/3.

www.thomann.de

Another stand design specifically dedicated to the iPad is the Z3 ($200, in black or silver) from Ratstands. This can be compared with a floor standing sheet music stand, and in fact, one of its major purposes is to display the iPad at a suitable height for reading sheet music scores or chord charts. It's great in other interior décor applications though offering convenient use of the iPad as a TV set, a recipe book in kitchens and so on.

FIGURE 5 Z3 stand

The Z3 folds flat but makes a bulky (20″ × 15″ × 2″) package weigh-
ing in at around 2.1 kg, since its base is intentionally quite heavy. The
telescopic shaft tilts back to the required angle and pulls upwards (with
heights from 16″ to 54″), and the iPad cradle allows landscape or portrait
use, and use with original model iPad or with iPad 2 using a small spacer.
The variable height and display angle offered by the Z3 make it suitable
for cellists, guitarists and in fact almost any musician in any application.
Using the iPad as a sheet music display, for viewing chord charts, lyrics

FIGURE 6 Mark Jenkins Z stand

or notes, as a sequencer, controller pad or guitar effects unit could all be made more convenient using the Z3, and the company also offers a 3 m charging cable so that the iPad can remain on charge while in its elevated position.

www.ratstands.com

SPEAKERS

The iPad, of course, has become a popular source of music playback and so there are many, many ways to amplify its audio output either through the headphone or through the dock connections. We won't look at any of these in detail, but will mention the move towards the wireless amplification using the iPad's Bluetooth facility, which is going to be relevant in a lot of live music performance applications. Bluetooth has a range of around 30 ft., so offers the possibility of quickly strewing speakers around a room ready for a live iPad-based performance without having to worry about actually wiring them up.

WoweeOne offers a Bluetooth enabled, rechargeable portable speaker which turns any surface it's placed on into a resonating surface, so it may be a handy way of amplifying an iPhone or iPad performance.

www.woweeone.com

Italian designed and US built, QbMito from SoundScience at around $100 is a small, cube-shaped speaker delivering a powerful Bluetooth-sourced 25 W of sound, running for 30 h on a charge and coming complete with a carry case. The more substantial Frankenspiel Panettone FS100 from the same company delivers 250 W wirelessly from left and right drivers and a front subwoofer, for around $500.

www.soundscience-corp.com

Of course, many other companies now offer wireless speakers. The Bose Soundlink at $300 has been well reviewed, as has the Audiovox Acoustic Research range which includes outdoor models which look like lamps or flower pots. The Sensia from Pure streams iPhone or iPad sound and also includes a DAB and FM radio.

KARAOKE

The iPad has found a place in Karaoke performances, thanks to its easy combination of touch control, audio and video playback. Karaoke systems vary from simply amplifying the iPad to integrating it much more closely into the performance.

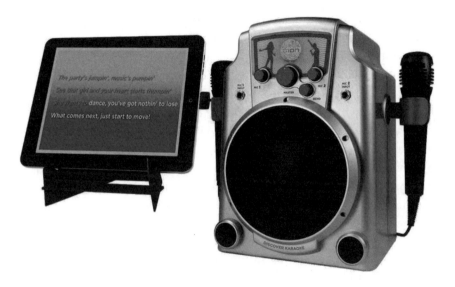

FIGURE 7   Ion Discover Karaoke

Ion offers the Discover system, which is a basic indoor Karaoke party setup with a built-in microphone accompanied by a free iPad app. Much more substantial is the same company's iPA46, more closely resembling a chunky keyboard combo amp, complete with a slot to mount an iPad on top.

www.ionaudio.com

Probably, the most powerful iPad-specific Karaoke hardware on the market is the Alto Professional iPa, which front-mounts an iPad or iPhone and offers a massive 400 W of amplification, speaker stand mounting, XLR/quarter inch inputs for additional mikes and an XLR output to slave more speakers. This system certainly doesn't look like a toy and in the United Kingdom shares distributors with Ion, Numark and Alesis.

www.altoproaudio.com

Apple stores also stock two-handheld wireless microphones for Karaoke applications both at around $100, the Soulo 1 and the Disney Spotlight. The latter includes pitch enhancement and effects and comes complete with 10 song backings from the Disney Channel – not from

Disney movie soundtracks – but presumably there's no limitation on what other songs can be loaded from iTunes, though they obviously won't have lyrics displayed. Soulo 1 also comes with 10 starter songs and allows a video to be shot of your performance.

An excellent source of Karaoke material for the iPad is www.ipad karaoke.net, which categorises songs either by title or by artist or you can search by style including R&B, New Wave, Musicals and so on. Tracks have a 30 s preview on the site and popular songs will exist in versions by several different companies, so you can have a choice of possible interpretations. Packs of 50 song backings in styles including Christmas, Country, Duets, Disco and many more can be bought for $50.

Karaoke apps for the iPad include Karaoke Anywhere; RedKaraoke which offers 45,000 songs with 20 free to start with; Glee Karaoke which offers 150 songs from the TV series; StarMaker, which is a 'Karaoke game with Auto-Tune'; Karaoke Now which supports backup to iCloud; a HipHop-oriented app PocketRap; a Bollywood Karaoke app; Karaoke apps designed for children including Twinkle Twinkle Little Star or Old McDonald HD and many others.

Obviously, there are many, many iPad apps for vocalists not specifically related to Karaoke performance, including VoiceJam, Vocalive, the Voix vocoder and many others, but that maybe the subject for another book.

HARD DISKS

The iPad isn't designed for use with external hard disk drives, and its data filing system uses private directories which make use with conventional disk directory systems difficult. You can use an app like FlashDriveHD or Jailbreak an iPad to overcome some of these limitations, but a more straightforward method now becoming common is to access a local hard disk using WiFi, which at least gives easy access to picture, sound and movie files.

Several small portable WiFi hard drives, of a size which make them not incompatible with iPad usage, are now coming onto the market. One which can be highly recommended is the Wi-Drive from Kingston

FIGURE 8 Kingston WiDrive

(16 GB $50, 32 GB $90), a tiny, slim device smaller than a pack of cigarettes which charges using USB (running for up to 4 h) and readily accepts movie, MP3 and image file transfers from your laptop. You can then load the free Wi-Drive app onto your iPad, go to Settings to detect and select the Wi-Drive's WiFi signature and can then instantly access files from the Wi-Drive.

Apart from effectively expanding your iPad's internal memory, this technique can act as a ready source of music as well as movie and image playback while avoiding dependence on programming all your content using iTunes. Up to three different users can access the drive simultaneously, all playing back different material.

www.kingston.com

Seagate have a competing model, the GoFlex Satellite 500 GB at around $175 featuring USB3 as well as USB2 and running for 7 h on a charge, streaming to up to eight different devices at the same time using its own WiFi facility but simultaneously allowing access to the Internet.

www.seagate.com

Maxell offers the stick-sized AirStash with a 16 GB memory card or larger and 7 h battery life but listing at $180.

www.maxell.com

Other models that are available include the Sanho HyperDrive iFlashDrive ($190), which has a dock connector rather than using WiFi and offers 32 GB storage, and the HyperDrive iPad hard drive which (once you've added your own 2.5″ SATA hard disk, or paid $400 for a model with 1 TB installed) connects to the iPad via a Camera Connection Kit without any need for jailbreaking – it also features SD card and Compact Flash card slots, so may be a handy option for transferring photos from a camera for viewing on an iPad as well as for applications playing back music or movie files.

Incidentally, for non-iPad specific hard drive purposes, in other words for backing up your iTunes File Transfer and other data from your laptop rather than directly from the iPad itself, it's hard to beat a portable drive such as the Western Digital My Passport. The 2 TB model weighs only a few ounces, supports USB3 as well as USB2, is powered directly from the laptop using a supplied cable and seems relentlessly dependable. When you do finally get home after performing, recording or shooting video with the iPad or iPhones, you'll find the My Passport, a reliable way to back up vast amounts of data – in fact, it's backing up this book as it's being written.

www.westerndigital.com

CHARGERS

If using an iPad for extended periods while making music, you may also like to look at charger options (unless your dock connector is already tied up). The iSound Portable Power Max at $70 offers five simultaneous USB/Apple Dock charging ports and will run an iPad for 30 h from its internal battery, or the same company's iSound4 at $20 will charge up to 4 USB/Apple Dock devices simultaneously. Alternatively, the Trent iCarrier at $100 or under offers 2 A- and 1 A power ports (one for a tablet and one for a smartphone) and doubles the running time of an iPad. In the United Kingdom, Maplin stores stock an excellent affordable four-output charger by Camelion which I currently use to ensure the performance readiness of my iPad concert setup comprising two iPads, an iPhone 4S and an iPod Nano.

CLOTHING

Before closing a discussion of accessories for the iPad and iPhone, it would be a shame not to mention at least one of the clothing lines which are coming onto the market with the intention of making all these devices more easily transportable. AyeGear based in Glasgow, Scotland, offers an excellent design in the AyeGear 22 jacket (£90 UK) or its sleeveless Vest version (£70 UK), both of which have a massive capacity for accommodating gear.

FIGURE 9 AyeGear Jacket

With 10-inch internal pockets, an iPad can easily be carried (in fact one on each side) and there's also an internal pocket for a smartphone with a transparent touch-sensitive cover so the phone can be used in place, in addition the cable runs to direct a headphone cable towards the user's head. There are pockets for coins and pens (with 22 pockets in all), a built-in key chain, zip pockets able to take a small camera, business cards or even a couple of CD's, and the jacket is fleece lined for warmth, nylon coated and machine washable.

Available in small to XXXL sizes, the AyeGear22 jacket has proven to be invaluable in many concert tours I've undertaken (see the Appendix for a list of equipment typically carried) – one of the many advantages when you're carrying large amounts of valuable gear being that you can take off the whole jacket at airport security searches and put it straight back on, rather than laboriously emptying your pockets of electronics.

www.ayegear.com

Chapter 9

Apps, WIST and Android

It's not the aim of this book to comprehensively review every music creation app available for the iPad – although that could be a subject for another
volume. At the time of writing, there are over 7,000 music apps available,
the majority also running (usually in some cut down form) on the iPhone.

The Apple Store's definition of a music app is extremely wide ranging. Many are simply playback devices for recorded music – effectively
new skins for the iTunes player. These do little more than re-arranging
or re-categorising your existing iTunes library, sometimes adding a visual
element to compensate for the fact that the iPad's basic iTunes app has
no Visualiser function. Popular apps include front panel simulations of
huge beatbox-style cassette players, reel-to-reel tape machines or in the
case of the cheekily named McIntosh AP1 Audio Player, a pair of illuminated moving needle VU meters, which are lovely to look at although
probably not very accurately calibrated.

Many other basic music apps – often free – are incredibly simple.
Some of the simple percussion instruments are quite handy and attractive – Bonang Barung is a simple Gamelan gong set and Chakra Chime
is a well-recorded Tibetan bell. There's a huge number of straightforward drum pad layouts offering percussion sounds, comedy noises, hits
useful for DJs or sound effects. Xcussion is a handy kit of Latin percussion sounds, BaDaBing is a basic and simply animated rock drum with
'bonus' bongos and Beat Pad Lite offers a simple set of electro drum
pads. Although all these percussion apps are easily played by hand,

don't expect any of them necessarily to interface to an external keyboard, MIDI or any other type of controller.

More complex drum machine apps such as the Korg iElectribe have been updated since their initial release to take advantage of the possibilities offered by MIDI interfacing. Korg's Wireless Sync-Start Technology (WIST) also now operates via Bluetooth to interface these apps to another iPad, iPhone or iPod touch. Simply enable WIST on each app and you're asked to choose which app is slave and which is master. Apps start and stop playing reliably together and stay in sync – although what happens if you change pattern or gradually change tempo is less predictable, but at least the system does offer the option of tempo synchronisation between two apps on different iOS devices without requiring full MIDI interfacing. At the time of writing, the apps able to use WIST include:

Korg iElectribe
Korg iKaossilator
Korg iMS-20
Korg Monotribe SyncControl
DM1 Drum Machine
Propellerhead ReBirth
Loopseque/Loopseque Mini
Anaphobia Mini
S4 Industrial Composer
Flail
Rhythm Studio
Yamaha TNR-i
BeatMaker 2
VirSyn iVoxel
VirSyn Addictive Synth
Arctic Keys
Tabletop
NLog MIDI Synth/Pro
MoDrum
Bassline

The WIST developer kit is available free for any iOS developers wishing to include WIST functionality in their designs at code.google. com/p/korg-wist-sdk.

Drum machine apps which do offer MIDI synchronisation include FunkBox from synthetic bits, which offers many options under the MIDI Connection Settings menu. Apart from choosing channel (traditionally set to 10 for drum machines), it's possible to switch MIDI clock transmission on or off, and there are various Latency settings which may solve problems particularly when the iPad is multi-tasking.

Other than percussion instrument apps, there are many very simple piano or organ iPad apps such as Nano Key Lite and keyboards with drum accompaniment or maybe electric guitar sounds, like Jam-Pad. Again these are unlikely to interface by either MIDI or any other method to external keyboards, controllers and other devices, and in most cases will have no method of saving performances – you'd have to use them either for very simple live performances or for recording onto a multi-tracker or other recording device.

More complex apps offer a virtual keyboard and some sequencer, accompaniment or drum machine facility. Among these is Akai Synth-Station, effectively an app version of the company's MPK series keyboard controllers. In SynthStation V3.0, you'll find a set of 9 drum pads, 9-channel mixing, a 2-octave virtual keyboard, XY expression pad, 3 triple-oscillator virtual analog synthesizers, 50 drum kits, a sequencer, arpeggiator, effects and filters. The app is core MIDI compatible which is handy because the virtual keyboard in the current version is extremely small, and the app gives many possibilities for creating dance-oriented or experimental beat-driven music.

SynthStation, of course, works seamlessly with Akai's own hardware controllers, the Akai SynthStation 25 for iPhone/iPod touch and the Akai SynthStation 49 for iPad, which we looked at in Chapter 6.

SampleTank from IK Multimedia, described as a 'sound and groove workstation', offers a dual virtual keyboard and 16 virtual drum pads, but it is effectively a sample playback device rather than a song composition facility. It loads sets of keyboard, instrumental, percussion and effects sounds and is closely dedicated to use with IKM's iRig MIDI interface (mentioned in Chapter 4) and so can help form an excellent studio or stage substitute for the now bulky-looking Akai, E-Mu and other hardware keyboard/samplers of the 1980s and 1990s.

FIGURE 1 Korg iMS20

Korg iMS20 synth app has an interesting history. In the 1970s, Korg launched the MS20 hardware monophonic analog synth, and the design was revived in the company's 'Legacy' software line, accompanied by an 84% size reproduction of the original synth in the form of a USB controller. Some time after the controller was discontinued, the iMS20 iPad synth app was launched and retains superb compatibility with the controller – if you can still find one. Using this compact controller with its minijack patch cables as the 'front end' of an iPad-based variable synthesizer can create a very compelling system, and the iMS20 app throws in a replication of Korg SQ-10 12-step sequencer (boosted to 16 steps) as well as additional synth parts, drum sounds played from pads, mixing and effects.

Because it's difficult now to find the Legacy hardware controller, it's more likely that you'll match this app with Korg's NanoKey miniature hardware controller – which works well using the Apple Camera Connection Kit – or with Korg's MicroKey25, 37 or 61, with an external power supply for the USB bus in the case of the 37 key model.

FIGURE 2 Korg Legacy MS20

OTHER APPS

Although this book doesn't aim to go into detail about the huge number of musical apps available for the iPad, it's worth mentioning a few outstanding examples to gain some idea of how much can be added to the potential performance of iPad apps by hardware controllers and other accessories. One prime example is Sunrizer, by the Polish app developers BeepStreet. This virtual analog synthesizer app was born as Horizon but had a name change early in its life and is roughly modelled on the Roland Jupiter JP8000 synth.

The JP8000 was an unusual design, offering a short four-octave keyboard (there's also a module version) and sound generation based on virtual analog modelling. This seemed a step back from the earlier JD800 synth which offered sampled waveforms matched with analog-style processing, and so, the JP8000 could never really compete in the creation of acoustic guitar sounds, choirs or anything other than the usual analog repertoire. However, it proved popular enough and found a great deal of use with all sorts of bands including Faithless, creating plucky analog sounds, pads, heavy bass sounds and so on.

FIGURE 3 Sunrizer

One unusual aspect of the JP8000 design was its inclusion of a SuperSaw waveform, effectively a number of sawtooth waves layered together for a thick, fuzzy sound. Apart from some cosmetic similarity in the panel design, the inclusion of an emulation of the SuperSaw waveform is a major feature relating Sunrizer to the original JP8000 design.

The similarities are by no means comprehensive though, and designer Jaroslaw Jacek has felt free to incorporate all sort of interesting design elements in the Sunrizer app. Just over two octaves of virtual keys are usually on display, but the keyboard slides up and down to access its full length. Pitch and modulation wheels are to the left, and the panel display offers two filters, two oscillators and two envelopes. Effects including delay and a highly programmable arpeggiator hide behind the small submenu displays.

Sunrizer's sounds are incredibly powerful, and I was pleased to be given the opportunity of programming some of the factory patches. The virtual analog sound creation system can be stretched to its utmost to create metallic, chaotic and digital sounding effects, and so Sunrizer can find use in almost any style of music.

Most significantly from the point of view of adding hardware controllers is the way Sunrizer's modulation works. Rather than simply adding vibrato or filter wow, the modulation control morphs between one Sunrizer sound and another which is completely independent. So, if A and B sounds are identical other than vibrato in the B version, using the modulation controller will simply add vibrato. But if the B version has completely different pitch, filter and effects settings as well as vibrato, all these will alter as the modulation wheel is used.

When compared with the modulation of synth sounds in Garage-Band – which doesn't even respond to Filter Cutoff messages – Sunrizer is more powerful, and so benefits much more from the addition of hardware keyboards and controllers.

The iSequence was developed also by BeepStreet some time before Sunrizer and is an eight-track sequencer app offering many built-in sounds. Since launch, these sounds have been expanded with optional packs, and now many of them are drum loops and patterns similar to those in the Apple Loops library. Although the iSequence sounds are relatively simple and don't have much response to external MIDI control, you can now set iSequence channels to send out note data to other MIDI modules. This offers the very exciting possibility of using the iSequence/iPad combination as a sequencer, either totally pre-programmed or running in a very live and interactive mode allowing notes to be punched in and out, and live transposition of patterns, controlling external MIDI keyboards, modules and samplers. This is a terrific combination for some types of music which leaves most hardware MIDI sequencers standing.

What you can also do with iSequence, in an improvement added some time after the original release, is to create your own sounds and loops with user sampling. Although this can be done surprisingly effectively with the iPad's built-in mike, this is where an audio input device such as Apogee Jam comes in.

I've done a great deal of sound and movie dialogue sampling with the Jam/iSequence combination, and the system works well.

FIGURE 4   iSequence

Once sampled, sounds can be pitch shifted, multi-split, enveloped, looped and processed with the internal filters and effects of iSequence. This is an incredible replication, in miniature, of systems which 10 years ago would have demanded whole racks of modules, MIDI interfaces, computers and audio processing devices.

ANDROID

Before concluding, we should take a brief look at other forms of tablet device running the Android or other operating systems to see whether they can offer any of the musical possibilities of the iPad. Although the iPad has dominated the tablet market, there are many superficially similar devices appearing from manufacturers including ASUS, Motorola, Archos and many others including generic models manufactured in the Far East, and many of these either approach the theoretical specification of the iPad or offer much lower prices right down to the level of $99.

At the time of writing, there's no mini-sized iPad (although this may change by the end of 2012), while many Android tablets are considerably smaller than the iPad, so the prospect of a music device which could fit into the pocket even more readily than an iPad is potentially very attractive. Certainly, Android tablets are perfectly capable as far as music playback and picture viewing are concerned (using the Android File Transfer app if you want to upload data from a Mac), so many users might find them an attractive possible alternative to the iPod or iPad.

Models like the Motorola Xoom Media Edition are specifically oriented towards downloading and playing back music and movies. The Xoom has excellent construction, long battery life, good sound from its headphone output and is a handy size for downloading and viewing

FIGURE 5 Motorola Xoom

movies. Other models such as the ASUS Transformer Pad TF300 have a removable keyboard, so the compact Android tablet has begun to straddle the worlds of netbook and laptop applications.

Even the briefest examination though of the Android Market (also known as Google Play), which sells the relevant apps, reveals a great discrepancy when compared with Apple's Apps Store. Although Apple offers around 7,000 music apps (in part because of all the iPhone development which went on before the iPad was even launched), the Android Market doesn't reach 700 at the time of writing, and the majority of these are simply music playback devices.

There's a fairly good choice though of simple drum and sound effects packages, some synthesizers and some audio recorders. These are becoming more and more ambitious each week, and it's certainly worth having a look at the Android Market to see what's appeared recently.

Simple recorders include the Hi-Q, which is a voice recorder which generates MP3 files, and there's also RecForge Pro which also records OGG files. J4T is a straightforward four-channel multi-track recorder, but a lot more ambitious is Audio Evolution Mobile, a multi-track audio recorder with a mixer and effects which has a professional looking multi-channel layout rarely matched by Apple apps.

There are various guitar tuners, and at least one voice pitch corrector Tune Me which aims to produce the usual auto-tune functions. In terms of musical instruments, there are several drum machines, including Electrum which also acts as a sampler, and RD3 Groovebox. Nanoloop and ReLoop are simple loop creation apps, and there are several virtual DJ turntables including Droid DJ. Dubstep music creation is well represented, among others by DubPro and Dubstep Universe.

There are versions of some familiar Apple app store residents including the MorphWiz theremin-like instrument, Reactable which is a complex sound creation modular system and Jasuto which also allows the user to match various modules to create abstract sounds. TouchDAW is one of several MIDI controllers which offers the possibility of using an Android tablet as a control surface for external instruments.

There are a couple of simple sound creation apps such as (imaginatively) Synthesizer, which offers simple sounds from a panel of touch buttons, Synthoid from Fleetway 76 or Sythe Free which creates sets of sine waves played from an XY controller.

Some of the Android synth apps use a simulation of FM synthesis. Not unexpectedly, these produce metallic and bell-like sounds vaguely reminiscent of Yamaha's DX-line synthesizers or the 1980s, so may or may not be to your musical taste. On the other hand, ADAM is a dual keyboard organ app with an overdrive distortion setting, and Blip Synthesizer is a matrix-based sound pattern generator like a very much simplified version of Yamaha's Tenori-On device.

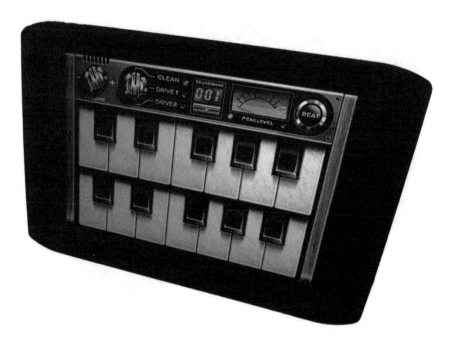

FIGURE 6 ADAM

There are some specifically imitative synths too – VL-Tone has an excellent screen simulation of Casio's original digital sound equipped calculator, although its sounds are not 100% convincing, but there is one very eye-catching line of synthesizer apps in the Android Market created by Nick Copeland of Bristol Audio software, comprising a series of simulations of classic synthesizers from the 1970s and 1980s. It's well worth visiting the Bristol Audio site to see the full range, originally developed for Linux operating systems. Nick Copeland has done an incredible job in simulating not only the sounds but also the panel layouts and cosmetics of synthesizers including the Minimoog, ARP Axxe, Moog Sonic 5 or 6, rare German BME 700, Oberheim OBX, ARP 2600 and many others. In the Android version, these cost a few pennies each and look fantastic when loaded on a device such as the Motorola Xoom Media Edition.

FIGURE 7 Minimoog on Motorola

Moog's Sonic 5/Sonic 6 are of particular interest to me as I owned the hardware original, performing and recording with one to the present day, including in a recent show at the British Museum in London. The Sonic 5 was developed in the 1960s by Musonics as a portable duophonic (two-note) analog synth with a built-in speaker and was revised by Moog into the Sonic 6 model with a slightly different casing. Neither model featured the original Moog 24 dB per octave lowpass filter, but the relative weakness of the filter compared with the Minimoog was more than compensated for by the addition of a ring modulator, dual LFOs which could be merged for complex modulation shapes, and various other facilities which made the Sonic 6 more versatile than the Minimoog as a creator of sound effects.

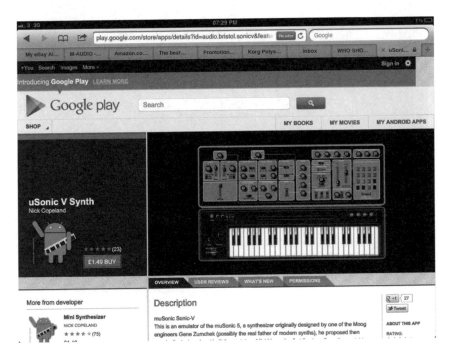

FIGURE 8   uSonic V

Copeland's Android app, the uSonicV synth, realises the Sonic 6 front panel beautifully, adding programmability and a digital delay to the original design. All the typical Sonic 6 features are there: double LFO modulation, LFO speed syncable to envelope decay, ring modulation for metallic sounds and much more.

Even more impressive is Bristol's ARP2600 simulation, which while resembling the original studio-oriented analog synth (used by Stevie Wonder, Edgar Winter, The Who and others) to a slightly lesser degree, squeezes in even more facilities including three oscillators, virtual patch cables and a digital sequence recorder which makes very complex sounding audio landscapes possible. These are about the most impressive musical apps and it's possible to run on an Android device at the moment.

FIGURE 9   ARP 2600

But the question remains, How well do they run? Unfortunately, the results aren't all that impressive. Android's screen operations have a very limited response, meaning that playing on virtual keyboards often suffers from unacceptable lag time. Playing sound effects on the uSonicV and ARP2600 is perfectly acceptable, but playing melodies is usually not. In addition to this, sounds particularly when using a delay or reverb tend to suffer from crackling and distortion, and it's usual for an app to crash entirely and to need re-installation.

To be fair, Nick Copeland advises against the use of his Android designs, far preferring the Linux originals, and other apps design companies have plainly stated that the delays inherent in the Android operating system at the time of writing have made it impossible for them to port over their existing guitar effects and other iPad apps to Android devices. Playing a guitar into an Android device and having the effected sound appearing maybe a quarter of a second later is never going to be musically acceptable.

The Android 4.1 operating system may deal with some of these delays, but without such improvements, it's difficult to imagine that this situation is going to improve any time soon, and so, Android and other tablet devices will be limited to music playback rather than music creation tasks for the foreseeable future. In addition, there's absolutely nothing in the way of music hardware which specifically interfaces an Android device for music creation purposes.

One possible forthcoming exception though is the Neiro from a new company Miselu.com, based in Soma in San Francisco, which is described as a 'portable, net-enabled social music device'. Neiro takes the form of a two-octave music keyboard mounted beneath an Android tablet and includes a Yamaha AudioEngine NSX-1 chip which handles the creation of synthesizer sounds and digital effects such as EQ, chorus and reverb. The company aims to include music apps from both Yamaha and Korg, and for example, there's a prototype version of the powerful sounding Korg PolySix analog synth in app form, which already exists as part of the Korg Legacy software package for the Mac and PC.

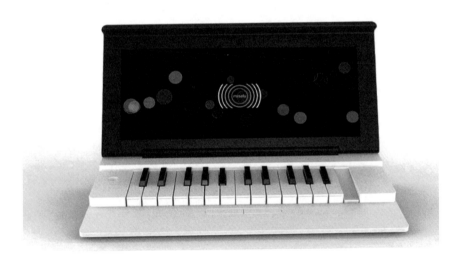

FIGURE 10 Neiro

Another app which already runs well on Neiro is N-Studio which includes a pad-based drum machine rather in the style of Roland's TR727, an XY pad sound modulator, eight-channel mixing and much more. Whether Neiro and other similar devices will bring the Android system into the world of music creation remains to be seen.

Conclusions

As successive versions appear, the iPad will only become more and more powerful and flexible, and its musical applications will become more and more diverse. At the time of writing, the main omission is still a powerful multichannel recording studio along the lines of Steinberg Cubase or Apple Logic. GarageBand on the iPad is an extremely limited substitute and is not receiving regular updates from Apple, and the other multichannel recorders that do exist don't offer built-in synthesizers or MIDI recording.

FIGURE 1 Auria

One powerful studio-style recorder Auria from www.Wave MachineLabs.com announced early in 2012 has so far failed to appear. It would offer 48-channel playback, very powerful mixer channels with eight subgroups, built-in VST effects (though each has to be modified by the designer before being compatible), convolution reverb, master EQ and compression. At the time of writing, Auria has supposedly achieved 18-channel simultaneous recording (though with what audio interface is unknown) and has reached the final review stage at the Apple Store.

FIGURE 2 Auria Recorder

If and when Auria appears, then the iPad will begin to rival many well-specified recording studios, hastening the process that began 20 years or so back with the invention of the PortaStudio and allowing musicians to work at home or on the move and become more and more independent of professional studios.

Certainly, the iPad is already contributing vastly to the move away from the recording studio and towards a very independent and mobile form of music production, and guitarists, keyboard players and singers as well as engineers and producers will come to rely more and more on this very powerful music creation device in the future.

Mark Jenkins
London, Summer 2012

FIGURE 3   M JENKINS

Appendices 1–7: Technical Specs, Studio and Stage Setup, Credits

APPENDIX 1: IPAD AUDIO CONNECTOR DIAGRAM

As discussed in Chapter 3, the iPad's audio socket is a four-pole mini-jack offering stereo out and mono in. The pin wiring is as follows:

Pin Number	Pin Name	Description
1	Tip	Left audio
2	Ring	Right audio
3	Ring	Common/Ground
4	Sleeve	Microphone

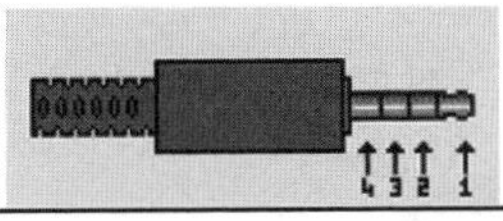

FIGURE 1 Connectors

Connecting a stereo headset or stereo line cable works because while it shorts out the microphone (pin 4) to the ground connection (pin 3), it will successfully connect with pins 1, 2 and 3 and allow normal sound output. Plugging in a microphone directly will not work; however, because it will short out pins 3 and 4 and fail to provide an independent signal to the microphone channel.

If these signals can be split using a simple adapter cable, then it is possible to use both input and output simultaneously. Speech Recognition Solutions, KV Connection and other companies make suitable adapter cables. Each splits the four connections to two separate jack sockets, so that the microphone in connects only to pins 3 and 4, and the other to pins 1, 2 and 3.

The microphone impedance must be over 800 Ohms to be recognised, and if so, it becomes the default sound source for audio recording and other apps, unless another microphone is detected connected to the dock port or via Bluetooth. A condenser microphone requiring power (such as many headset mikes used for telecoms) will not work, although small in-line adapters are also available, which can supply power to these as required and which cost only a few dollars.

FIGURE 2   Splitter

APPENDIX 2: IPAD 30-PIN DOCK CONNECTOR DIAGRAM

Apple's proprietary 30-pin dock connector is used on all the iOS devices including the iPad and has changed slightly in its implementation since its introduction. At the time of writing, there's talk of replacing the connector altogether, but for the moment anyone contemplating designing for the connector could refer to the pinout chart below:

Pin	Signal	Description
1	GND	Ground (–), internally connected with pin 2 on iPod motherboard
2	GND	Audio and Video ground (–), internally connected with pin 1 on motherboard
3	Right	Line Out R (+) (audio output, right channel)
4	Left	Line Out L (+) (audio output, left channel)
5	Right In	Line In R (+)
6	Left In	Line In L (+)
7	?	
8	Video Out	Composite video output or Component Video Pb
9	S-Video Chrominance output	or Component Video Y
10	S-Video Luminance output	or Component Video Pr
11	AUDIO SW	If connected to GND device sends audio signals through pins 3–4, otherwise uses onboard speaker.
12	Tx	Sending line, Serial T $\times$ D
13	Rx	Receiving line, Serial R $\times$ D
14	RSVD	Reserved
15	GND	Ground (–), internally connected with pin 16 on motherboard

(Continued)

Pin	Signal	Description
16	GND	USB GND (–), internally connected with pin 15 on motherboard
17	RSVD	Reserved
18	3.3V	3.3V Power (+) stepped up to provide +5 VDC to USB on Camera Connector Kit.
19,20	+12V	Firewire Power 12 VDC (+)
21	Accessory Indicator/ Serial enable	Different resistances indicate accessory type: 1 kOhm – docking station, can take some devices into photo import mode. 6.8 k – Serial port mode. Pins 11-13 are TTL level. 68 kOhm – makes iPhone 3g send audio through line out. 500 kOhm – related to serial communication. 1 MOhm – Belkin auto adaptor, some devices shut down automatically when power disconnected.
22	TPA (–)	FireWire Data TPA (–)
23	5 VDC (+)	USB Power 5 VDC (+)
24	TPA (+)	FireWire Data TPA (+)
25	Data (–)	USB Data (–)
26	TPB (–)	FireWire Data TPB (–)
27	Data (+)	USB Data (+) – pins 25 and 27 may be used in various ways, to force or drain charging.
28	TPB (+)	FireWire Data TPB (+)
29,30	GND	FireWire Ground (–)

The back side of the dock connector is wired as follows:

2 4 6 8 10 12 14 16 18 20 22 24 26 28 30
1 3 5 7 9 11 13 15 17 19 21 23 25 27 29
Pins 1&2, 15&16, 19&20 and 29&30 are connected on the motherboard.

Obviously, experimentation with the iPad dock connector is not to be recommended. Simply knowing that pins 1, 5 and 6 access stereo line level audio input is not going to be sufficient to achieve anything significant, whereas using serial data transmission/reception on pins 12 and 13 would require a thorough knowledge of Apple serial protocol. However, increasing numbers of companies are going to the effort of investigating this in a systematic manner to create Apple-approved products – see details in Appendix 3.

APPENDIX 3: IPAD IOS DEVELOPMENT

If you'd like to develop apps for the iPhone and iPad, you can join up as an Apple developer here–

https://developer.apple.com

More detailed information on the SDK or Software Development Kit is here–

https://developer.apple.com/ipad/sdk/

Any apps developed for iOS are submitted for examination and sale through the Apps Store, with a percentage of sales going to Apple.

Anyone wanting to develop hardware which interfaces electrically with iOS devices (cases and stands for example being exempt) need to join up to the hardware development program here–

https://developer.apple.com/programs/mfi/

This Mfi licensing program specifies the hardware components, tools, documentation, technical support and certification logos needed to create AirPlay audio accessories and any electronic accessories that connect to iPod, iPhone and iPad.

You can't get feedback or advice from Apple regarding a potential Mfi development product before joining the Mfi scheme, nor find out about the royalty rates which will be levied until a credit check and contract are completed.

As a hobbyist intending to develop an Mfi accessory for personal use and not for mass production, you won't be able to join the Mfi program. Apple recommends you 'use a third-party solution which will allow you to connect iOS devices to serial devices and to write iOS apps that communicate with these serial devices'.

APPENDIX 4: MIDI MESSAGES

The iPad and iPhone support Core MIDI and the MIDI specification have been well defined for over 25 years. GarageBand and other apps should handle MIDI data from keyboards and other types of controller, but this may be limited in implementation. Other than Note On/Off messages with velocity, here are some controller messages recognised by GarageBand.

> Controller CC1 – Modulation
> Controller CC7 – Volume for current track (in instrument mode, not in arrangement mode)
> Controller CC8 – Pan for current track (in instrument mode, not in arrangement mode)
> Controller CC64 – Sustain Pedal
> Channel Aftertouch
> Pitch Bend

Not implemented in GarageBand are the controllers usually used for Filter Cutoff and Resonance for synthesizer apps.

Key mappings for drums are as follows:

Classic Studio, Vintage and Live Rock Kits:

> C1 – Bass Drum
> F1 – Low Tom
> C2 – Mid Tom
> A1 – High Tom
> D1 – Snare Drum
> C#1 – Side Stick
> D#1 or E1 – Rimshot
> Bb1 – Open HiHat
> F#1 – Closed HiHat
> G#1 – Pedal HiHat
> C#1 – Crash Cymbal
> D#2 – Ride Cymbal
> F2 – Ride Cymbal Bell

MIDI Play, Stop and Timecode messages are not supported by GarageBand.

Program Change messages are not recognised.

No MIDI is sent out.

Virtual controllers in the user interface, including mod wheel, do not always update in response to MIDI messages.

APPENDIX 5: IOS-COMPATIBLE MIDI DEVICES

This is a list at the time of writing of MIDI instruments which work with iOS devices, noting in some cases whether connection through the Camera Connection Kit is sufficient or if a powered USB hub is needed:

Akai EWI wind instrument USB

Akai LPD8 pad controller

Akai APC-40 (via powered USB hub)

Akai LPK25 (has power issues for some users)

Akai MPK49 and MPK61 (via powered USB hub)

Akai MPK25 (via USB hub)

Akai MPK Mini

Akai RPM3 (audio only)

Akai Synthstation (works via USB hub)

Akai Synthstation 49

Alesis DM6

Alesis DM10

Alesis iO Dock

Alesis iO2 (older model, via USB hub)

Alesis Photon X25

Alesis Q25

Alesis Q49 (works via USB bus – no external power necessary)

Arturia Analog Factory controller keyboard

Audiotrak MIDI Mate

American Audio VMS4 DJ (works with power supply, audio on channel 1 works also)

Arturia Analog Factory 'The Laboratory' 49 Key (using supplied power adaptor)

Axiom 49 (2nd Gen)

Axon AX 50

Behringer BCF2000/BCR 2000 control surfaces
Behringer iAxe 323 USB Guitar (audio only)
Behringer UMA25S (audio and MIDI)
Behringer UMX25
Behringer UMX61
Behringer UMX490
Bespeco MIDI Cable
Carillon M1X1
Casio DG-20 Guitar Synth
Casio Privia PX-130
Casio WK7500
CME U2MIDI
CME U-Key 49 Mobiletone
Dave Smith Instruments Tetra
Digitech GNX4 (audio and MIDI)
DSI Mopho Keyboard
Doepfer Dark Energy
Doepfer Dark Time
E-MU X MIDI 1 × 1
E-MU Xmidi 2 × 2
EDIROL PCR-M80
EDIROL PCR-1 (via USB Hub, audio interface also works)
EDIROL PCR-A30 (via adaptor or USB Hub, audio doesn't work)
EDIROL PCR-30
EDIROL PCR-300
EDIROL PCR-800
EDIROL UA-1D (via USB hub, adds noticeable latency)
EDIROL UA-25EX (via USB hub, audio only)
EDIROL UM-1ex
EDIROL UM-1sx
EDIROL UM-2ex (Roland UM-2 version does not work)
Elektron TM-1
ESI KeyControl 25 XT
ESI M8U
ESI Midimate II
Evolution MK-361C
Focusrite VRM Box (via USB Hub, hardware only)

Genovation 900-MPC Midi Patch Changer
HDE USB to MIDI Cable
JamHub TourBus (via USB hub)
Kawai MP8II
Korg KP3 KAOSS PAD
Korg Kaossilator Pro
Korg K25
Korg Kontrol 49 (with power adaptor)
Korg microSAMPLER and microKONTROL
Korg M3
Korg M50
Korg microKEY 37 (via powered USB hub)
Korg MicroX
Korg MS-20 (via hub)
Korg NanoKey, NanoPad and NanoControl
Korg Pandora PX5D
Korg R3 (using supplied power supply)
Korg RADIUS
Korg Triton TR Series
Korg X50
Kurzweil PC3
Livid Block (via powered USB hub)
Livid Block Ohm64 (via powered USB hub)
Lexicon Omega
Line 6 Pocket POD (MIDI only)
LogiLink USB MIDI Cable
Logitech Premium Notebook USB Headset
Mackie XD-2
Mistar MidiLink
M-Audio Axiom 49 (1st Gen)
M-Audio Axiom Pro 61
M-Audio Evolution X-Session
M-Audio Keystudio
M-Audio O2
M-Audio Oxygen 8
M-Audio Oxygen 8 (3rd Gen)
M-Audio Oxygen 25 (3rd Gen)

M-Audio Oxygen 49 (via powered hub)
M-Audio Oxygen 61
M-Audio Prokeys Sono 61
M-Audio Prokeys 88
M-Audio Prokeys 88SX
M-Audio Uno 1x1 (older Midiman models do not work).
M-Audio Keystation 49a
M-Audio Keystation Pro-88
M-Audio KeyRig
M-Audio XSession Pro
M-Audio Venom
Moog Little Phatty Stage II
Moog Multipedal
Novation Remote 25 LE
Novation Remote SL series
Novation Remote Zero SL MK1 (via hub or battery)
Novation X-Station (with power adaptor)
Novation Xio 25 (audio and MIDI)
Presonus Audiobox USB (audio and MIDI, via powered hub)
Prodipe Midi USB 1i1o
Roland A-300PRO, A-500PRO, A-800PRO
Roland Cakewalk A-500S Master Keyboard (with batteries)
Roland FP-7F
Roland HP 305
Roland Lucina
Roland Octapad SPD-30
Sequentix Cirklon

Sonuus i2M Musicport – the company also offers B2M and G2m monophonic bass and guitar to MIDI convertors which presumably also work.

Studiologic VMK-161 (via powered hub)
Studiologic VMK-188
Swissonic MIDI-USB 1 × 1
Terrasoniq MIDI ONE
Terratec Axon AX 50
Turtle Beach USB MIDI 1 × 1 (via powered hub)

V-Machine VST Player
Waldorf Blofeld (desktop version)
Yamaha E430
Yamaha Ez250i
Yamaha KX-8
Yamaha PSR-S900
Yamaha S90ES
Yamaha Stage Piano 3P00 (external power needed) Yamaha UX-16
Yamaha WX-5 EWI (with adaptor, via USB hub)
Yamaha YDP-V240
Yamaha YPG-235
Yamaha YPG-625
Zoom A2.1u (audio only, via USB hub)
Zoom H4n (audio only)
Zoom R16 (audio only, via USB hub)
The following devices do not work:
Access Virus TI2
Akai MPD16
Akai Synthstation (works via USB hub)
CME UF8
EDIROL UA-20
EDIROL SC-8850
EDIROL PCR-A30
EDIROL UM-1SX
Emagic MT-4
Emagic Unitor MkII
E-MU Xboard 49
Evolution UC33
Hercules DJ Console MK1
ICON-Global-Midiport
Line 6 POD HD200
Line 6 POD XTLive
Line 6 POD XT Live
Korg Micro X
M-Audio Axiom 25 (1st Gen)
M-Audio Axiom 49 (1st Gen) (mixed reports)
M-Audio iControl

M-Audio Midisport 4 × 4
M-Audio Radium 61
Midiman Midisport 4 × 4
M-Audio Oxygen V2
M-Audio Oxygen 8 (1st Gen)
MidiMan MIDISport 2 × 2
MOTU MIDI Express 128
Native Instruments Traktor S4 Kontrol
Native Instruments Maschine
Novation Launchpad
Novation Remote 25/37 (may need firmware update)
Roland JUNO-G
Roland SH-201
Roland UM-1S
Roland UM-ONE
Vox JamVox
Yamaha MO8

APPENDIX 6: MARK JENKINS

iPad Concert and Studio Hardware Setup 2012
www.markjenkins.co

All products mentioned are used and recommended by the author.

Stage;
AyeGear jacket (dual iPad plus iPhone pockets)
Cygnett Aerosphere iPad sleeves
iPad 64GB 1: IKM iKlip stand, Apple Camera Connection Kit,
    iSmart roll-up USB keyboard
iPad 64GB 2: IKM iKlip stand, Korg NanoKey or remote MIDI
    keyboard + Line 6 MIDI Mobilizer II
iPhone 4S
Korg Mini Kaoss Pad
Belkin Rockstar signal combiner
Ear Pollution Ronin DJ Headphones

MARK JENKINS
at the Vanderbilt Planetarium, Long Island, New York July 2011
www.markjenkins.co

FIGURE 3   M JENKINS VANDERBILT

Studio:
  Alesis i/ODock
  TASCAM iU2 interface
  Korg Microstation synthesizer
  Western Digital My Passport hard disk
  Kingston Wi-Drive wifi hard disk

Apps:
  Beep Street iSequence (sampling sequencer)
  Beep Street Sunrizer (synthesizer)
  Korg iMS20 (lead line synthesizer)
  Korg iKaossillator (sound effects)
  Korg iElectribe (drums)
  Synthetic Bits FunkBox (drums)
  Xcussion (latin percussion)
  IK Multimedia SampleTank (sampled sounds)
  Fairlight (sampled sounds)

Voix (vocoder)
IKM VocaLive (vocal effects)
MorphWiz (sound effects)
Thicket Classic (sound effects)
AirScratch HD (dj app)

plus two hundred others ...

APPENDIX 7: BOOK CREDITS

Special thanks to–

Alquimia (equipment and support)
Anaïs Wheeler (Taylor & Francis)
Kattie Washington (Focal Publishing)
Leila Stocker (Turtle PR)

and to

Paul Kaufman (IK Multimedia)
Jeff Lea (Western Digital)
Helen Marriott Smith (Gadget Show Live)
Jaroslaw Jacek (Beep Street music apps)
Phil Smith, Brad Delava (Sonic Distribution/Apogee)
Maggie Zaboura (Kingston/Zaboura PR)
Jon Bickle, Donna Ward-Smith (Numark, Akai)
Christian Basener (Line 6)
Kerry Hogarth (Cygnett)
Neil Wells (TASCAM)
Dave Fanning (Tin Drum PR/Digital Seasons)
Zain Sehgal (AyeGear)
Adam King (Asus)
Billy Burnett (Motorola)
Maxine Rushworth (Jackie Cooper PR)
Anna Taylor (Griffin/Talk PR)
Bill Fox, Howard Moscowitz, Dale Parsons, Dave Sneed (iPad concert
support in the USA).